W9-ABY-173

The Adman in the Parlor

The Adman in the Parlor

Magazines and the Gendering of
Consumer Culture, 1880s to 1910s

ELLEN GRUBER GARVEY

New York   Oxford
OXFORD UNIVERSITY PRESS
1996

Oxford University Press

Oxford   New York
Athens   Auckland   Bangkok   Bombay
Calcutta   Cape Town   Dar es Salaam   Delhi
Florence   Hong Kong   Istanbul   Karachi
Kuala Lumpur   Madras   Madrid   Melbourne
Mexico City   Nairobi   Paris   Singapore
Taipei   Tokyo   Toronto

and associated companies in
Berlin   Ibadan

Copyright © 1996 by Ellen Garvey

Published by Oxford University Press, Inc.,
198 Madison Avenue, New York, New York 10016

Library of Congress Cataloging-in-Publication Data
Garvey, Ellen Gruber.
The adman in the parlor : magazines and the gendering of consumer culture,
1880s to 1910s / Ellen Gruber Garvey.
p.   cm.
ISBN 0-19-509296-1; ISBN 0-19-510822-1
1. American fiction—19th century—History and criticism.
2. Short stories—Publishing—United States—History—19th century.
3. Periodicals, Publishing of—Economic aspects—United States.
4. Popular literature—United States—History and criticism.
5. American fiction—20th century—History and criticism.
6. Short stories, American—History and criticism.
7. Literature and society—United States—History.
8. Advertising, Magazine—United States—History.
9. Books and reading—United States—History.
10. Women consumers—United States—Attitudes.   I. Title.
PS374.S5G34   1996
813'.409—dc20   95-9467

1 3 5 7 9 8 6 4 2
Printed in the United States of America
on acid-free paper

For Janet
and in memory of my parents,
Robert Garvey and Miriam Gruber Garvey

Acknowledgments

A magazine's masthead reminds readers of a few of the many people who must be involved in causing it to appear, but a book has only the author's name on it and hides the long list of people whose help has, whether intentionally or unknowingly, been behind it. I would like to thank the people whose time, attention, and patient and generous readings at many stages of this book's production enriched it.

At the University of Pennsylvania, I am very fortunate to have had Peter Conn in on this project from its earliest stages, with unfailing encouragement and a lightly guiding hand.

The present and former members of my writing group contributed years of wise comments and smart remarks, willingness to read drafts in varying stages of dishabille, and the flexible ability to both productively nitpick and see the larger picture. I am grateful to Ilana Abramovitch, Jane Holzka, Harriet Jackson, Ellie Kellman, Nancy Robertson, Nina Warnke, and Vera Whisman. They have been the community of readers every writer needs, a welcoming and demanding audience and a source of deadlines.

Other thinkers whose conversations, challenges, and questions have made their way here include Peter Stallybrass, Janice Radway, Richard Ohmann, Susan Greenfield, Roland Marchand, Betsy Erkkila, Valerie Traub, and Susan Zlotnick. Catherine Nickerson and Susan Williams each read through an earlier draft of the manuscript and made pertinent overall critiques. For comments on portions of the manuscript I wish to thank, among others, Rachel Kranz, Sally Mitchell, Regina Kunzel, and Joan Arnold. Barbara Rusch, Diane DeBlois, and other ephemerists have generously passed information across the divide between collectors and academics. For less easily defined but equally necessary support and encouragement on this project I am grateful to Rita Barnes, Lydia Buechler, Myra Goldberg, Janet Golden, Howard Gruber, Kim Hall, Steve Messina, Miles Orvell, Barbara Katz Rothman, Eric Schneider, Ellie Siegel, Suzanne Slater, Nancy Stiefel, and Doris Wallace. Tom Tuthill generously helped photograph ads, and Ann Garvey and Nancy Stearns, along with others already named, helped proofread. Bert and Ellen Denker allowed me the use of their scrapbooks.

At Oxford University Press, I would like to credit Liz Maguire and Elda Rotor for their contributions; Cynthia Garver's and Ann Kraybill's copyediting sharpened and improved the manuscript.

Two fellowships at the Winterthur Museum and Library were crucial to my work on trade cards, and allowed me to discover and explore trade card scrapbooks. The erudite staff of the library, Neville Thompson, Jill Hobgood, E. Richard McKinstry, Heather Clewell, Gail Stanislaw, Shirley Giresinger, Kathryn Coyle, and Bert Denker were unfailingly helpful and gracious. Katharine Martinez, Patricia Elliot, and Dot Wiggins helped make Winterthur such a productive place to conduct research. At the Strong Museum, I had the advantage of the knowledgeable help of Deborah Smith, who also read an early version of the scrapbook chapter.

The Rutgers Center for Historical Analysis extended important resources to me for my research. I would like to thank the members of the RCHA seminar on consumption, whose conversation and questioning, under the leadership of Rudi Bell and Victoria de Grazia, deepened my thinking. I am particularly grateful to Kathy Peiss, Jackson Lears, Joe Broderick, Kathleen Hulser, and Erica Rappaport for their helpful comments.

Janet Gallagher's unfailing enthusiasm for this project, her willingness to tolerate the material culturing of piles of paper in inconvenient locations, and her knack for laughing in the right places have been almost as crucial as her love.

Contents

The Adman in the Parlor

Each month the magazines appear;
  Their clever covers may be seen
Upon the stands twelve times a year,—
  And weekly papers in between.
  They tell about things Phillipine,
Or war 'tween capital and wages;
  But on those things I am not keen,
I like the advertising pages.
  . . .

I like to read of gorgeous gear,
  Of latest brands of Breakfastine;
Of Peerless Household Near-Veneer,
  Of magic stuff to scour and clean,
  Of Boneless Beef and Bakeless Bean,
Of bargains for all sects and ages;
  From steamboats to a soup-tureen,
I like the advertising pages.

Carolyn Wells, "A Ballade of Advertisements,"
Profitable Advertising, December 1904, p. 766

Introduction

Magazine: 1. A place where goods are laid up; a storehouse or repository for goods and merchandise. . . . 5. a. Used in the titles of books, with the sense (fig. from 1 . . .): a storehouse of information on a specified subject or for a particular class of persons   b. A periodical publication containing articles by various writers; chiefly a periodical publication intended for general rather than learned or professional readers, and consisting of a miscellany of critical and descriptive articles, essays, works of fiction, etc.

Oxford English Dictionary

A successful magazine is exactly like a successful store: it must keep its wares constantly fresh and varied to attract the eye and hold the patronage of its customers.

Edward Bok, editor, *Ladies' Home Journal*[1]

The older definition of magazine—a repository for goods and merchandise—from which the meaning of a periodical containing a miscellany of articles derives, survives in French and other languages. The *grand magasin* of the nineteenth century was that new and exciting institution, the department store, where goods were arrayed in visually exciting, sensual displays within architecturally novel buildings and where new plate glass windows brought the displays onto the street, creating new habits of looking. Especially for middle-class urban women, shopping became a new form of entertainment: a shopper could easily spend an entire day in the department store, moving from its concert hall, to its ladies' parlor to write letters and meet friends, to its art gallery, to its various goods for sale.[2]

The *grand magasin* reframed the task of shopping as a luxurious and eminently pleasurable pursuit. Similarly, the magazine joined in one package, or booklet, the commercial world of goods and sales with the world of private musing and romantic fantasy. This book examines how stories work within the magazine, in which, as in the department store, all elements participate in selling.[3] Rather than providing a comprehensive overview of the magazines and advertising of this period, this book analyzes points of interaction between fiction and advertising to learn both about fiction and about advertising's place in the culture. Magazines are texts embedded within the world of commerce and the world of their readers. Read this way, magazines no longer appear to be the site of a war between commerce and culture, in which literary or editorial interests are separate from and in conflict with advertising and commerce. Rather, we discover that advertising and fiction acted on one another in complex and unexpected ways. The advertising-supported magazine as an institution has buttressed the interests of advertisers and the commercial discourse as a whole, and constructed the reader—especially the female reader—as a consumer. Even as advertisements became touchstones of modernity and its fragmentations, ads came to seem natural and ordinary to readers at the end of the nineteenth century. The larger question this book asks is, how did this happen?

Frank Munsey, editor and publisher of one of the first new magazines oriented to the middle class in the 1890s, asserted that "fiction . . . is responsible for enormous circulations, and without fiction the general advertiser would find the magazine proposition quite a different matter and decidedly uninteresting from a business standpoint." The editors of the advertising trade journal *Profitable Advertising* concurred: "Magazines are undoubtedly read chiefly for the stories, and it is therefore evident that the storywriter is one of the advertiser's most valuable assistants."[4] Stories, like the department store's musical concerts, were part of what drew readers to the magazines, making them potential buyers. But advertisers also depended on stories to create a climate in which their ads would persuade readers to become buyers. Ads, as the most overt representatives of the magazine's commercial basis, had a complicated and varied relationship to its stories. Ads provided glimpses of life that were excluded from stories, and opportunities for pleasure and play more accessible than those that stories offered; at times ads also depended on stories to accustom the reader to their techniques and concerns. Any insistence on an editorial/advertising split distorts the experience of actual magazine readers, who took in a magazine as a whole.[5] Some 1890s commentators claimed that readers specifically sought out the advertising, and found as much value in it as in the editorial matter. When we read them together, we find advertisers learning from fiction writers, while fiction writers define themselves both within and against advertising. The reader is invited to move between the two.

Magazine ads, initially isolated in their own discrete sections at the front and back of each issue of middle class and elite magazines, were brought into text columns in a very few magazines in the 1890s; it was not until the late 1910s that significant numbers of magazines mixed ads among articles and stories throughout, ultimately creating the magazine form we are familiar with today, with stories chopped up to continue on nonconsecutive pages. Just as in the late-nineteenth-century department store, where a shopper might shift focus from one display to another or see them all simultaneously while pursuing his or her own business, in magazines that did mix ads in with the reading matter, a magazine reader began to find that "concentration on any one thing is segmented, and distraction is a natural and pleasing element. . . . Concentration or attention [gives] way to a moment-by-moment multiple focus."[6]

Even before most magazine layouts fragmented narrative and information by breaking up stories to "tail" onto the advertising pages, the magazine drew the attention of readers to the ads, inviting them to shift continually among the pages, using the "moment-by-moment multiple focus" they were already familiar with as department store shoppers. By joining many activities in a single bound unit, the magazine invited the reader to interrupt reading a story about a marriage proposal to consider how she would look in an attractive jacket or how useful her husband would find a mail-order course on law, or to envision her children contented and healthy from eating Quaker oatmeal.

This fragmented or multiple focus in both two- and three-dimensional space was elicited in many venues besides the magazine and department store in the 1890s—mail-order catalogs, fairs, expositions, the new museums, and even children's scrapbooks were a few examples. All were metonyms of one another; all swapped techniques for display and organization among themselves; and each helped naturalize one another for their viewers or readers.[7]

Fiction and advertising within the magazines were similarly enmeshed. It has become a critical commonplace that there is no pure sphere of literature from which fiction emerges, untouched by the commercial nexus within which it is published and within which its writers live and work; instead, fiction constantly if uneasily reflects on its place within commerce. But relatively little attention has been given to other elements in this reciprocating relationship. Advertising, too, reacts to the fiction around it—fiction that has prepared both readers and writers to imaginatively enter into scenarios made attractive and familiar by advertising.

The Chapters in Brief

Readers' interaction with advertising has never been a passive process of absorbing advertising messages. Chapter 1, "Readers Read Advertising into Their Lives: The Trade Card Scrapbook," examines ways that one largely female group of readers actively played with advertising, by looking at traces of that play in 1880s scrapbooks of advertising materials. Through this interaction, these readers were constructed, and constructed themselves, as consumers.

Middle-class girls and adolescents made scrapbooks of trade cards, a colorful and attractive advertising medium for nationally distributed products of the 1880s. These were forerunners of the explosion of the widespread magazine advertising for such goods in the 1890s, and were largely displaced by it. As they compiled their books, the girls learned to be responsive to advertising and to become consumers in the new arena of mass-produced, nationally distributed products. The scrapbooks both record this learning process and show the compilers at work creating meaning from their arrangements. Compilers practiced applying "taste" to create unique assemblages out of mass-produced objects and learned to express themselves through such play.

The collector-compilers of trade card scrapbooks typically mixed the trade cards, with varying degrees of differentiation, with calling cards, reward of merit cards, holiday greeting cards, and Sunday school proverb and psalm cards. The formal similarities between commodity advertising and religious cards sanctioned advertising. They made it all the easier for the girls to enthusiastically integrate the commercial world with the social, school, and religious worlds. Manipulating ads also offered possibilities of transgressive play, using materials not valued or closely supervised by adults. As these children learned to fantasize in the language of advertising, the consumer arena seemed to offer itself as a sphere of free play and pleasure.

Beyond issuing trade cards, advertisers initially did little to actively encourage this play with advertising. By the 1890s, however, as magazines became economically dependent on advertising rather than on sales of copies or subscriptions, magazine publishers, acting in the interests of advertisers, developed an institutional interest in focusing the attention of readers on advertising. Chapter 2, "Training the Reader's Attention: Advertising Contests," looks at one strategy magazines pursued both to assure advertisers that their ads were being read and to demonstrate to readers that ads were entertaining, informative, and worth reading. Through the contests, magazines helped advertisers as a group by structuring readers' imaginative interaction with advertising in the interests of constructing them as consumers.

Because ads ran in the back of most magazines, separated from the material for which readers presumably bought the magazine, advertisers might assume that readers skipped over their ads. Magazines presented the contests to potential advertisers as an incentive to advertise, a warranty that their ads would be read, and often as an index of how well they were read.

Contests encouraged readers to see the ads as an important part of the magazine, and even to feel obligated to read them. Readers of children's magazines in particular often joined clubs and through other activities asserted their membership in a community of magazine readers. They were encouraged to transfer that sense of participation in the community of the magazine to loyalty to advertisers, and to help support, and thereby create, the magazine through their actions as consumers.

Just as the formal similarities between trade cards and religious cards helped to legitimize the pleasures of playing with advertising, contests, too, appropriated familiar pastimes and adapted them to commercial purposes, inviting readers to apply the old forms to new material. Advertising contests encouraged readers to bring advertising materials into their lives and to incorporate brand names and advertising slogans into their conversation and writing. Advertising figures became their companions, and advertising, perhaps as much as the magazine's fiction, became a reliable source of friendly, attractive characters.

As advertising increasingly entered daily life, writers and commentators grappled with the meaning of the new advertising discourse. Chapter 3, "'The

Commercial Spirit Has Entered In': Speech, Fiction, and Advertising," tracks the more diffuse movement of advertising into national culture. Brand names and ad slogans were useful as a common frame of reference in an increasingly heterogeneous country. As the national distribution and advertising of goods by brand name shaped a national vocabulary, the cultural shorthand they created enabled people across the United States to understand a reference to a brand of soap or a joke about an advertising slogan. Allusions to articles in even the most intimate use within the individual and private world of the home took on a new kind of public accessibility.

While some commentators were critical of what they saw as the invasion of advertising into unsullied space, others cited what they saw as the benefits of advertising to people at large, separate from any use advertising might have for the product's potential consumer. These views tended to divide along class lines: commentators were more likely to disapprove of advertising or references to it that appeared in elite sites and situations than of advertising or advertising references in nonelite settings.

One arguably elite setting in which advertising references began to appear was fiction. Often such references were satiric, but in a popular novel of 1888, Amelie Rives's *The Quick or the Dead?*, brand-name references are more central. Rives's novel embodied questions about the individuality of people and relationships, and the duplicability and replaceability of relationships, in part through references to brand-named items associated with characters. I argue that the novel attempted to reconcile the idea of individuality and irreplaceability with the system of mass production in which all duplicated articles are equally authentic.

Beyond the specifics of brand names, magazine fiction found other ways to demonstrate the multiple desires and social meanings embodied in the purchase of a product. Chapter 4, "Reframing the Bicycle: Magazines and Scorching Women," takes up the question of how advertising and fiction interacted in relation to a single commodity. When the safety bicycle in the 1890s made bicycling accessible to women, wheelwomen found themselves riding through contested terrain. The new mobility that bicycles offered was both attractive to feminists and the target of attack by conservative forces. Both defense and attack took medicalized form, with pro-bicyclers asserting that bicycling would strengthen women's bodies for motherhood, and anti-bicyclers claiming that riding would not only masculinize women, but would ruin their sexual health by promoting masturbation. Bicycles were attacked as a danger both to women's sexual purity and to their gender definition.

The discourse of consumption constituted by the fiction, articles, and advertising within the advertising-dependent middle-class magazine of the 1890s subsumed both feminist and conservative views in the interest of sales. Bicycles were often advertised in the middle-class magazines of the period, constituting 10 percent of national advertising in the 1890s. Individual manufacturers took the masturbation threat literally and addressed it by promoting the sale of crotchless "hygienic" saddles, but could not counter the real threat for which female bicycle masturbation was a metaphor: women on the loose, mobile and independent. Doing what individual advertisers could not do for themselves, magazines acted in the interests of advertisers in the aggregate and published numerous bicycling stories that recontained the threat posed by the bicycle's offer of freedom. A counterpressure, however, made itself felt in works published in non-ad dependent magazines, where the potential subversiveness of women's bicycling was celebrated rather than recontained or muted.

But a woman didn't buy a bicycle every day. Chapter 5, "Rewriting Mrs. Consumer: Class, Gender and Consumption," looks at a larger pattern of changes in women's relationship to commodities and their purchase and finds advertising, stories, and the institution of the magazine itself addressing and constructing women readers as consumers.

As advertisers increasingly defined women as their target audience, advertising-dependent magazines presented their women readers with fiction that encouraged them in their role as consumers. This encouragement took different forms depending on the class of women addressed. Magazines addressed to cash-poor women presented ways to earn money to buy advertised goods and helped to justify their purchase, while suggesting that such consumption could be consistent with their values of thrift and moral responsibility. Magazines addressed to middle-class women, on the other hand, discouraged autonomous work for married women and encouraged them to seek fulfillment in shopping and the emotional caretaking of their families. These magazines valorized the apparent power available to women as shoppers through courtship stories that were allegories of shopping, and which featured women choosing wisely between offered choices.

Middle-class married women were steered further away from earning opportunities as writers by the magazines' firmer espousal of professionalism. While some women's magazines had allowed readers to make less formal forays into the literary market, the newer magazines reduced such opportunities.

Chapter 6, "Men Who Advertise: Ad Readers and Ad Writers" explores the status of the ad reader—increasingly presumed to be female—and the ad writer, presumed male. While women were being positioned as advertising readers, "good writing" of both ads and fiction in the middle-class magazine was being defined as writing for men and often by men: concise writing for the busy businessman. Advertisers and the magazines that acted for them expressed their ambivalence about both addressing women and publishing them. Women were to be advertising readers and consumers, not writers.

When commentators discussed whether ads were valuable to readers, their conclusions were mixed. But when they discussed whether ads had value for women, their writings were far less equivocal. Women were said to seek education and entertainment from the ads, turning to the advertising pages first, in preference to the reading matter, and hanging onto every word of the advertiser, grateful to be entertained. Ad trade journals published their own versions of the courtship-shopping stories, in which admen used their skills to woo a mate.

Background: The Changing Magazine

The big change in magazine economics has usually been pinned to 1893, the year that three monthlies—*Munsey's*, *McClure's*, and the *Cosmopolitan*— dropped their prices to ten cents, shifted the basis of their enterprise from sales to advertising, and began to achieve circulations in the hundreds of thousands.[8] However, ten-cent and even five-cent monthlies that depended on advertising for their revenues and had circulations of half a million to one million had existed before this. Many in this earlier group, called mail-order monthlies, such as the *People's Literary Companion*, *Ladies' World*, *Comfort*, and *Youth's Companion*, were addressed to a poorer, more rural readership. They were about the size of a tabloid newspaper and had ads mixed on the page with stories and other material. Others, such as the *Ladies' Home Journal*, which grew out of this group and aligned itself with the new middle-class ten-cent magazines, saw their audience as a genteel one. But the three new ten-cent middle-class monthlies of the 1890s, at about seven by ten inches, with ads restricted to the front and back pages, more closely resembled the form of a third group: the older elite magazines such as *Harper's*, *The Atlantic Monthly*, and *The Century*.[9] And yet, as Richard Ohmann has documented, the new ten-cent magazines were different in content and audience from the older elite magazines; as they achieved such large circulations, they necessarily reached readers who had not previously subscribed to magazines.[10] Aside from the mail-order magazines, in 1885 there had been only four general monthlies in the United States with a circulation of 100,000 or more; their combined circulation was 600,000. By 1905 twenty general monthlies had such circulations, and they shared an aggregate circulation of over 5.5 million.[11]

Price was an important reason that ten-cent magazines reached readers who hadn't subscribed to the "quality class" or elite publications *Harper's*, *The Atlantic Monthly*, and *The Century* at thirty-five cents a copy/four dollars a year, or *Scribner's* at twenty-five cents a copy.[12] But the new middle-class ten-cent magazines took advantage of new technologies and economies of scale to offer more illustrations and livelier layouts than the elite magazines. Their interest in commerce and industry extended beyond the advertising pages into editorial selections as well. It had been the job of Gilded Age quality class editors "to sift, scrutinize, and select literary manuscripts, always watching over established boundaries of taste and propriety."[13] These genteel editors did not actively plan the magazine but expected unsolicited manuscripts to drop in over the transom to the "Editor's Study," or "Editor's Easy Chair," as column titles in two of the elite magazines named the editor's workplace. (An 1892 advertising trade journal satirized this tendency to dozy passivity, referring to the "Editor's Spare Room," "Editor's Folding Bed," and "Editor's Easy Socks."[14]) In the new magazines of the 1890s, however, editors moved from a supine to an upright posture; the new editors actively solicited and commissioned articles and stories, and shaped the magazine. Frank Munsey, for example, editor and publisher of *Munsey's*, who prided himself on his wide-awake entrepreneurship, wrote his column from the "Editor's Desk." Often these editorial solicitations were inspired by the needs of the advertisers.

The new readership cultivated by the middle-class ten-cent magazines, if only because of its size, was a less homogeneous group than the mainly North-eastern elite readers of the "quality class" publications. To some, the appearance of this new readership signaled the end of an era, and prompted changes in the old-line magazines as well. Looking back, L. Frank Tooker, an editor at *The Century*, complained of the instability of this new readership:

> There was the problem of a new group of readers so numerous that it gave tremendous weight to the sort of thing that it was eager to devour. Yet to label the reading of this body was difficult; it was catholic, but erratic. . . .
>
> Hosts of new magazines had arisen to make greater the problem—syndi-cated magazines; muck-raking magazines; art magazines; the magazine of spe-cial interests, like the tired business man and the woman who wanted her home to be beautiful. . . . And most of these were cheap, and could be thrown aside after hasty reading, like a newspaper. And with all this mad clutter, we were being left behind. . . . We stood as one among many, distinguished only, with a few oth-ers, by our dignity.[15]

Less querulous magazinists saw new opportunities in the new magazine readership. Brander Matthews was a prolific contributor to magazines of all types whose opportunities to publish expanded with the new, often higher-pay-ing magazines. His 1893 "Story of a Story," provides an allegory of the reach of the new magazines. It follows the adventures of an inspirational story as it cir-culates from its dying author to the editor of the *Metropolis* (evidently based on the *Cosmopolitan*, where Matthews had been a staff contributor since 1890).[16] The *Metropolis* editor sees the story as something to help "balance the mid-summer fiction"; from him it goes to a well-heeled buyer who reads it on a train to the country, where his discussion of it with the woman he loves gives him an opening to propose. Next, the story goes to an elderly war widow who is reminded by it of her youth; then to a bigamous blacksmith on a western Indian reservation, who tells the visiting engineers who have brought it that it has "done [him] good to read" (they promise to tell the editor of the magazine, the cousin of one of the engineers); and finally to the son of the illiterate scrub-woman who retrieves the magazine from the railroad car she's cleaning. The already virtuous lad finds it "a splendid story!" and, infused with the desire to model himself on the hero, goes to work in a lawyer's office where "he devel-oped true manliness, energy, character, and ability" and becomes first a lawyer and then an honest assembly representative.[17]

The theory of the circulation of magazines embodied in Matthews's story proposed not only that lasting effects could spring from the seemingly ephemeral magazine but also that these periodicals had a life extending well beyond their month on the newsstands. In Matthews's story, the *Metropolis* reaches both elite readers (such as the well-heeled train passenger) who might have read the "quality class" publications, and typical mail-order monthly read-ers such as the Western blacksmith and the scrubwoman's son. But as

Matthews's version of the new magazine spreads out through the class ranks and from the metropolis into the West, it disseminates middle-class values and even a reforming spirit.[18] Its editor, after all, is cousin to an engineer, the pro-totypical forward-thinking managerial man.

The most crucial distinction between the new ten-cent magazine and the older elite magazines, and the distinction on which the other differences rest, was the reliance of the new magazines on advertising rather than sales, with advertising rates pegged to circulation figures. With this change, publishers made a definitive shift—from selling magazines directly to readers to selling their readership to advertisers. Yet even as editors and publishers reoriented their obligations and accountability from readers to advertisers, it became more important to create magazines that appealed to readers, since a larger reader-ship made magazines more attractive to advertisers. The lower prices brought in that larger readership, principally composed, Ohmann suggests, of the new professional managerial class of the city and suburbs: a group that still excluded rural and working-class readers.[19] Although both the new middle-class ten-cent magazines and the cheap mail-order magazines, addressed to a poorer, often rural audience, depended on advertising rather than sales, their different readerships, along with their respective spending power, shaped distinctive ways of promoting the interests of the advertisers as a whole. One result of this was a substantial contrast in their treatment of women's work.

Important differences persisted between the old "quality class" magazines, the new ten-cent magazines, and the mail-order magazines beyond the 1890s. But since the "quality class" publications became increasingly interested in attracting advertising, differences in this respect diminished. This book will focus less on the differences between the groups of magazines than on the ways in which advertising came to shape magazines—the ways in which advertising-dependent magazines acted in the interests of advertisers in the aggregate, and the ways in which magazines as a group constructed the reader as consumer.

Advertising was not just a highly visible element of the new magazine but something akin to its bindery stitching as well. The fact that many different companies advertised in the magazines would suggest that no individual or class of advertisers could exercise any great power or control over the maga-zines. In a well-publicized move in 1893, Cyrus Curtis and Edward Bok of the *Ladies' Home Journal* had excluded advertising for patent medicines and finan-cial schemes from their magazine without incurring disaster.[20] By putting a lit-eral warranty on the products advertised in their pages, they actually made their ad space more desirable.[21] Other magazines boasted of their freedom from advertising influence. Beyond these anecdotal and self-promoting moments, however, which reassured readers of the reliability of both editorial and adver-tising matter, magazines came to act in the interests of advertisers in the aggre-gate, and advertising came to shape magazines.

Sometimes it is possible to document advertising's shaping of magazines in direct terms. Mark Sullivan, an editor at *Collier's*, and earlier at the *Ladies' Home Journal*, recalled in his memoirs a request from Charles William Eliot, the president of Harvard, for help in excluding automobiles from Maine's Mount Desert Island. Sullivan told him the project was doomed: "Newspapers and periodicals could not help him; the economic basis upon which they existed was advertising, and at that time about half of all advertising was of automobiles and accessories."[22] More recently, Gloria Steinem, introducing the first advertising-free issue of *Ms.* magazine, revealed much about the pressures women's magazines are under to "supply what the ad world euphemistically describes as 'supportive editorial atmosphere' or 'complementary copy,'" requirements codified in the advertising insertion orders. Such orders stipulate placement of a company's ads adjacent to related editorial matter, meaning that the magazine must therefore either carry editorial matter on fashions or home furnishings or lose clothing and furnishings advertisers. Advertisers' insistence on publishing product-related copy is overwhelming. According to one "beauty writer," magazines keep advertisers based on how many editorial clippings mentioning the product they produce per month.[23]

Editors of such 1890s ten-cent monthlies as *Munsey's* might not have been disturbed by Sullivan's or Steinem's accounts. They asserted in editorials that they saw no conflict between editorial and advertising interests, not because business interests would be kept separate from editorial matters, but because advertising itself would be valuable to the reader, and because the reader and advertiser had interests in common.

Advertising became increasingly important to magazines that were not part of the flowering of the ten-cent magazine as well. Though some kept a high cover price and never achieved the high circulations of the ten-cent magazines, many sought other ways to demonstrate their value to the advertisers. *St. Nicholas*, at twenty-five cents a copy, had a far smaller circulation at the beginning of the twentieth century than several other children's magazines, but offered advertisers entree into more affluent households. As more magazines competed for advertisers, *St. Nicholas* was one of the magazines that identified its task as the creation of a congenial environment for ads. Along with *Ladies' World* and *Good Housekeeping*, it took up this task overtly, by establishing contests that directed the readers' attention to their advertisements. These magazines thus provided advertisers with what amounted to an early version of survey results that reassured advertisers their ads were being read, while they encouraged readers to see ads as a source of pleasure and play.

Munsey's, notable as the first of the middle-class magazines to drop its price to a dime, articulated its mission as bringing fiction and ad writing closer to one another, thereby training readers through its fiction to appreciate the directness and succinctness of advertising writing. Frank Munsey's editorials explicitly framed his preferences in fiction in terms of terseness, action, and force—and then praised advertisements for embodying these virtues:

> The modern advertisement is a thing of art, a poem, a sledge hammer, an argument—a whole volume compressed into a sentence. Some of the cleverest writing—the most painstaking, subtle work turned out by literary men today—can be found in the advertising pages of a first rate magazine. Every word is measured, examined under a magnifying glass, to see just how big it is, just how much meaning it has, and how many kinds of meaning it has.[24]

The virtue of brevity and compression here is its efficiency: conveying more in less time, like the labor-saving devices advertised in the back of the magazine for wide-awake readers.

Progressive Advertising

Other editors and publishers went ahead and created a congenial environment for ads without overtly identifying that as a task for themselves. Advertising and magazine professionals developed new ways to discuss and think about advertising. Defining oneself as an advertising professional claimed distance from the earlier model of advertising embodied in medicine shows, the entertaining hyperbole of P. T. Barnum, and other shady if amusing frauds of what one writer calls "the advertising game."[25] The new advertising agencies and copywriters attempted to align advertising with modernity. They suggested that advertising was valuable not only because it would increase business but also because it would gain the advertiser respect. It would be a sign that the advertiser was up to date and, in a favorite term of the period, *progressive*. The political reformers who came to be labeled progressive shared with business the ideals of efficiency, the exaltation of "scientific" approaches, and respect for experts. But the term as it appeared in business discourse did not register political alignments. Rather it described business practices in tune with what was cast as progress. The word *progressive* was part of the title of at least five trade magazines of the 1890s, from *Progressive Bee-Keeper* to *Progressive Printer*. And it was specifically linked with the new emphasis on advertising as a respectable, anything-but-fraudulent activity.[26]

The ten-cent advertising-supported magazine fulfilled the increasing need of manufacturers of nationally distributed brand-name goods for media in which they could advertise nationally. Such manufacturers in the 1880s had put colorful advertising trade cards into the hands of thousands, but nationally circulated magazines were clearly a more efficient tool. The national advertisers that flocked to these magazines sometimes touted new commodities like bicycles, snapshot cameras, and typewriters, or new forms of familiar goods, like centrally processed and nationally distributed grains. Others promoted goods like scouring soaps and crackers, no longer sold in bulk, but given a trademarked identity and sold nationwide.[27]

Advertising trade writing used the term "progressive" to flatter businessmen who envisioned national markets for their goods, and retailers who embraced new modes of display and cooperated with the manufacturers of nationally distributed goods to sell their products. The term was used circularly: progressive businessmen advertised, therefore one could recognize a progressive businessman by his advertising. The term identified as well such newer business practices as selling magazines below the cost of production and finding the profit elsewhere in the transaction. Editors of the new magazines expressed their sense of economic alignment with business and advertising through the word "progressive," asserting that advertising educated readers and improved their lives. As the editors saw it, advertisers, readers, and editors were all on the same side. As Bok put it, "The making of a modern magazine is a business proposition; the editor is there to make it pay. He can do this only if he is of service to his readers, and that depends on his ability to obtain a class of material essentially the best of its kind."[28] In this case, the material that was the best of its kind included fiction. Advertisers depended on stories not only to make the magazines attractive to readers but also to create a climate in which their ads would succeed in persuading readers to become buyers. Editors therefore had reason to prefer stories that neither detracted from the premises of advertising nor undercut the tone of the ads, but rather created fictional worlds as much as possible continuous with the concerns of the ads.

The conventions of what was characterized by readers and editors of the time as "realist" fiction naturalized, or accustomed readers to expect in fiction, a universe of domestic minutiae, a densely described fictional middle-class world of finely calibrated, socially significant detail, in which objects through metonymy or synecdoche stand in for their owners. In these stories, possessions are inventoried, and purchases and desired purchases are laid out to create a fictional reality or tell the reader of the characters' longings. In William Dean Howells's *A Hazard of New Fortunes*, a novel much concerned with the magazine business and serialized in *Harper's Weekly*, the March family practices possession-reading skills as they search for a New York apartment. They judge the character of each prospective landlord or sublessor by "reading" his or her possessions as they survey the rooms.

Late-twentieth-century critics have drawn a tighter circle around the field of realism, chiefly treating such writers as Howells, Dreiser, Wharton, and James as its proponents, and pointing to their close description of social relations as characteristic of this realism, or, more recently, discussing realism as a "strategy for imagining and managing the threats of social change."[29] So the Marches both read and reflect on the act of possession reading as they perform it at the home of Basil March's new employer, the nouveau riche magazine patron. What if the possessions don't actually reflect the possessor's taste?

> The drawing room . . . was delicately decorated in white and gold, and furnished in a sort of extravagant good taste; there was nothing to object to the satin furniture, the pale soft rich carpet, the pictures, and the bronze and china bric-a-brac, except that their costliness was too evident; everything in the room meant money too plainly, and too much of it. The Marches recognized this in . . . hoarse whispers . . . ; they conjectured from what they had heard of the Dryfooses that this tasteful luxury in no wise expressed their civilization. "Though when you come to think of that," said March, "I don't know that [their landlady's] gimcrackery expresses ours."[30]

But this more complex and critical approach of Howells, James, and Wharton appeared chiefly in the more elite magazines. Stories in the ten-cent magazines more often celebrated the sense that the reader inhabited the same world as the characters and would be pleased to recognize references to its novelties and familiar characteristics. Advertisements depended precisely on the possession-reading skills that the Marches both exhibit and critique, on alertness to the world of domestic minutiae, and on acceptance of its realist codes, promulgated in ever more elaborate detail. In this respect, the ads and fiction in the magazines were in continuity. The kind of detail that this realism depended on, realism's attention to ownership as an index of taste and thus of character, was closely tied to the world of advertising and to the assumptions that governed ads. Brand names became a short-hand version of this type of realism. When they appeared in fiction, they quickly indicated to readers that the characters inhabited the same universe as their own; when advertisements deployed familiar fictional formulas, brand names had much the same effect.[31]

According to advertising historians, advertising between the 1890s and 1910s was dominated by rational appeals that sought to persuade chiefly by attracting readers' attention and informing them that goods were available. In this chronology, advertising argumentation does not sprout an associational dimension until the late 1910s, when graphics make goods "resonate with qualities desired by consumers—status, glamour, reduction of anxiety, happy families—as the social motivations for consumption."[32] Similarly, it wasn't until the 1920s that product mentions were woven into complex verbal narratives, as radio ads and their characters brought the social context of consumption into the advertising strategy.[33]

But complicated narratives that embedded products in a social context and associated them with romance, happiness, freedom, social acceptance, and socially approved behavior did appear in the 1890s—in fiction that seemed to serve no commercial purpose. We might think of such fiction as preparing the place that advertising would eventually take up: it prefigured later strategies applied more deliberately by advertisers, while it trained readers to appreciate and respond to the kinds of narratives that advertising came to use. Product-focused stories did not endorse individual brands, but in effect promoted product categories, along with the general sense that shopping choices and consumption itself were important.

The mechanism of the advertising-dependent magazine harnessed this capacity in the interests of advertisers and made magazine fiction a selling tool in a larger sense. Formulaic stories in the magazines flattered women readers' power as shoppers; these stories valorized the readers' ability to choose wisely among products as well as among suitors. In a sense, the magazine itself, like the department store, exalted consumerist choice making. Both magazines and department stores were considered respectable spaces in which middle-class women could wander; unlike a city street, the department store was an enclosed, monitored space. A visit there would not excite comment, and once inside the shopper was allowed seemingly limitless choices. The magazine, too, was monitored space, from which questionable, risqué, or politically incendiary material was excluded; within it the reader could wander seemingly at will, choosing among articles, stories, poems, and ads.[34]

1

Readers Read Advertising into Their Lives: The Trade Card Scrapbook

The taste for collecting is one that grows by what it feeds upon.

Everyone possesses, or has possessed, a scrap-book of some kind or other. . . . What histories lie hidden between the leaves—true histories of our inmost selves—each addition to our scrap-book a mile-stone on our mental journey!

both from Janet Ruutz-Rees, *Home Occupations*, 1883[1]

I did not believe that a class of girls could reveal such possibilities in a heap of advertising cards.

Letter from a missionary in India,
thanking a Sunday school class for their scrapbooks[2]

How did nineteenth-century readers read and understand advertising? Although the advertising industry has always been interested in knowing whether its ads attract buyers, market research data doesn't tell us what advertising meant in the lives of readers. But we can learn how readers in the 1880s and 1890s saw and interacted with advertising, and sometimes even literally what they made of it. As mass-circulation advertising-supported magazines spread in the 1890s, they became the major medium of national advertising. But the readers of these magazines had already learned to interact with national advertising through another widely distributed medium: the colorful advertising trade card of the 1880s. Readers left behind traces of their inter-action in collections of advertising trade cards arranged in scrapbooks. Play with advertising materials offered the arrangers, probably mainly children and adolescent girls, an attractive new mode of self-expression and expertise, even

16

as it trained them as future consumers and naturalized the developing discourse of advertising. It allowed them both to express and to imaginatively construct a new relationship between the national commercial culture and the individual home.

Girls structured interaction with advertising along the lines of cultural practices that were already familiar to them, such as making quilts and compiling autograph books. In interacting with advertising, they both demonstrated and extended gendered expertise in such areas as attention to visual aesthetics of composition and gradations of color. By treating religious and commercial material similarly, both advertisers and girls wrapped a measure of cultural approval around the newly developing pleasures of shopping.

The cards established a niche that was soon to be filled by magazine advertising. That is, although magazines of the 1870s and 1880s already contained advertising, the trade cards helped construct a particular orientation toward it, the sense that it was a space of pleasure, fantasy, and free play—an orientation that magazines went on to foster and channel by promoting new kinds of play with advertising, as we will see in chapter 2.

Consumption

From about the 1880s on, the move from home production to buying goods in the marketplace, which had been under way for centuries, shifted ground.[3] Now middle-class urban women, in particular, increasingly bought goods that had once been produced in the home; they replaced the locally made generic goods that had already replaced home production with brand-named nationally distributed goods. So while such women had bought soap for decades, they now switched from buying rough blocks cut by the grocer to buying molded packaged soap sold under a brand name, which they had learned of through its national advertising. Similarly, while hand milling of grain was rare even in a rural household by this time, consumers now not only bought milled and processed grain in bulk, but learned of brand-named packaged flour, oats, and processed cereals through national advertising. Shopping became a more central part of middle-class urban women's work, with shopping itself increasingly often a matter of choosing among brands and patronizing the new institutions of shopping: the grocery store that sold a growing proportion of brand-named goods, and the elaborate department store. New cultural practices grew up around shopping: it became possible to spend an entire day in a department store. These cathedral-like "palaces of consumption" offered customers not just floors of elaborately and theatrically displayed goods, but spaces that were ostensibly entirely social: ladies' lunchrooms, ladies' parlors for writing letters and meeting friends, concert halls, and art galleries.[4] Grocery stores displayed shelves of packaged goods, with colorful labels repeating the figures and logos familiar from advertising. The visual profusion on the shelves both suggested that there was an abundance of goods (even if the rows of packages all contained soap, or all flour) and announced that there was an abundance of *kinds* of soap, *types* of flour, and that one must learn to distinguish between them via their packaging. Consumers learned that shopping entailed negotiating visual complexity.

Middle-class urban women increasingly understood the world and their relation to it through a discourse of consumption. Unlike the classes below, the late-nineteenth-century middle class had money to participate in the new kinds of shopping; a middle-class family might, for example, have the economic lee-way to choose packaged goods over cheaper bulk goods. Unlike the classes above, it had a large enough membership to shape mass institutions. For mid-dle-class families, consumption became a way to actively articulate their class position; they consolidated class definition by establishing taste markers, and then by choosing appropriately among them. Learning to shop as a middle-class woman, then, expressed both gender and class position.

The education of the shopper also included inculcation into newly impor-tant modes of advertising, especially those used to advertise the mass-produced and brand-named goods that were increasingly significant in the marketplace. The vocabulary of brand names became more central, and advertising forms developed to cue consumers to this new vocabulary. One of these new forms was the brightly colored advertising *trade card*. Beginning in the 1880s, trade cards dominated advertising for nationally distributed products, until they were largely supplanted by national magazine advertising during the 1890s. Both arenas of advertising contributed to a national culture in which people all over the country could recognize a reference to an advertising slogan or brand of stove polish.

Readers of 1880s advertising cards left traces of their readings in scrap-books, in which they grouped the cards in ways that presumably pleased or made sense to them. The historical position of the young children and adoles-cents who compiled such books gives these scrapbooks a special status; they provide a window into readers' responses to the magazines that soon became central in organizing a culture of consumption. As young people, scrapbook compilers were apprentice shoppers, entering the discourse of consumption and learning to converse within it and respond to its conventions and impera-tives. Additionally, as literate middle-class children and adolescents specifically of the 1880s and early 1890s, they were of the same class and culture as the readers of the ad-dependent magazines of the 1890s and 1900s. Magazine ads thus took the place of trade cards within the production of advertising, as well as within the imaginations of a generation of readers primed to appreciate them. The advertising readers who made trade card scrapbooks both responded to the kinds of accretionary, miscellaneous forms they were familiar with, such as newspapers, catalogs, department stores, exhibitions, and maga-zines, in which commercial and noncommercial messages mingle, and pre-pared themselves for a world in which advertisements were ever more present.

Making scrapbooks from tradecards was widespread; while the over sixty scrapbooks examined for this study were mainly from the Eastern states, such scrapbooks were also made in Indiana, Illinois, Ohio, Michigan, and Wiscon-sin, and probably further west as well.[5]

The Trade Card

Advertising histories written by or drawn from accounts by early advertising agents locate the roots of magazine advertising in newspaper advertising. This version of advertising evolution is logical from the perspective of advertising agents, who usually sold space first in local newspapers, then in combinations of local papers, and finally in nationally distributed periodicals.[6] But from the point of view both of producers of nationally available merchandise, and of the advertisement readers who were their potential customers, nationally distributed trade cards were at least as important a precursor to magazine advertising. "At the height of its popularity in the 1880s, the trade card was truly the most ubiquitous form of advertising in America," the historian Robert Jay asserts.[7]

Trade cards often operated at the intersection of the national manufacturer and the retailer. Typically 2 x 4- or 3 x 5-inch cards, trade cards were imprinted with the name and often address of a retailer or manufacturer and a picture, sometimes related to the product, sometimes not. As they created recognition of brand-name products, they helped to override habits of shopping in which articles were requested generically, or, in this period before the self-service grocery, in which the shopper relied on the grocer to choose or recommend what to buy. Manufacturer's trade cards sometimes explicitly suggested that customers scold or refuse to buy from grocers who did not carry their product; others told the consumer to be on the lookout for retailers trying to pass off an "inferior" product. They suggested, in other words, that the customer had a closer relationship with a distant manufacturer than with the familiar grocer. Other trade cards countered the facelessness of the distant advertiser by assigning it a character and positioning the company or its representatives almost as members of the family: for example, the cards used to advertise Lydia Pinkham's vegetable compound were designed to resemble the sepia-toned carte-de-visite photos exchanged socially (see figure 1-1).[8]

Cards were nationally distributed by the drummers or salesmen who sold merchandise to storekeepers, who in turn bestowed the cards on their customers.[9] Such cards benefited both the manufacturer and retailer. To promote brand recognition, "manufacturers were eager to supply the local merchant with whatever advertising materials he might agree to use in dealing with his customers. . . . The generous supply of chromolithographed trade cards and other advertising materials was in fact a powerful incentive for the general store owner to agree to stock the product in question."[10] Cards for nationally distributed goods were even more valuable to the retailer when they advertised the local store as well—often with a chromolithographed ad for the product on the front, and the name of a retailer stamped or printed in letterpress either on the back or in a front border. At least one soap company distributed colorful cards with its name on the front and ledger columns printed on the back, handy for a grocer to use in writing out the receipt that would go home with the groceries.[11] Cards also came as inserts with some packaged products, such as soap and coffee.

PUT THIS IN YOUR ALBUM.

Figure 1-1   This trade card for Lydia Pinkham's
Vegetable Compound mimics carte-de-visit
cards and positions Pinkham as an intimate.

Chromolithography and the Value of Color

Though trade cards for stores had existed since the 1700s, their profusion and
popularity as an advertising device was a product of the refinement of chro-
molithographic printing, enabling cheap, high-volume reproduction of fluidly
designed pictures in color.[12] Engravings, another popular form of pictorial
reproduction, had necessarily been printed in one or at most two colors, and
any other application of color was a hand process. Although hand coloring
could be done assembly line fashion, it still made an item in color both relatively expensive and unique. When cheap chromolithography became available
in the late 1860s, advertisers were drawn to the attractions of color in the new
chromolithographs. An 1890s advertising reference advises, "The mechanical
picture of a parlor stove, even though presented with the most carefully drawn
surroundings, in black and white cannot have the ruddy glow of indoor cheerfulness, as will the same stove pictured in a run with color."[13]

Trade cards arrived on the scene alongside mass production and mass reproduction: every card an identical duplicate, advertising products that were themselves identical, mass-produced artifacts. As I will discuss, the identical-ness of the trade cards was itself often a valued attribute for collectors.

Trade cards were only one of several kinds of chromolithographed pic-tures. High-art oil paintings and watercolors were reproduced by the same process. These, however, were soon the target of elite scorn, in part because the democratized availability of an image and its broad dissemination were seen as subtracting value from the original, in terms not so different from Walter Ben-jamin's objection to the loss of the "aura" of the original in later photolitho-graphic reproduction of a painting.[14] Although chromolithographed reproduc-tions of paintings, often called chromos, could refine the home and encourage a love for the beautiful, *Literary World* complained that "Under the present sys-tem of wholesale gratuitous distribution, their office will be degraded and they will rank as Sunday School picture cards."[15] Even before the 1880s, chro-molithography's cheapness and liberal use of color, along with its free distri-bution in trade cards and as premiums, had made chromolithographed repro-duction both a medium and an entire genre that was sneered at by adherents of high art standards. But those who used and collected chromo cards recorded no ambivalence about the mass reproduction of art. In fact, the cards' mass pro-duction was what made them available for play. And the cards did not claim to reproduce an already existing artifact. As with other mass-produced objects, each was the same, no one card more original or authentic than another. They had thus not lost the "aura" of the original, for in a sense, there was no original; conversely, through identical reproduction, each copy was equally original. Trade cards therefore offered their recipients the chance to play within the interstices of mass production and individual taste.

While some cards were lithographed specifically for a manufacturer or retailer, others used generic or stock pictures.[16] Often the same chromo design—perhaps of flowers or birds—was overprinted with a name to make calling cards or name cards for individuals, with biblical verse to make the religious cards used in Sunday schools that one printer called proverb and psalm cards, or with an advertising message to make trade cards for retailers[17] (see figure 1-2).

Late-twentieth-century eyes may find the printing gaudy and crude. Nonelite nineteenth-century viewers, however, found the color enormously alluring. Color was a gift to its recipients, an enticement to keep the card. The attraction of chromolithographed cards is dramatized in an 1881 episode in Laura Ingalls Wilder's autobiographical novel *Little Town on the Prairie*, in which a new arrival introduces the name card fad to the Dakota territory town girls. Wilder's loving description of the cards recalls her own sensual appreciation of the novelty of chromolithography in the 1880s: "The cards were palest green, and on each was a picture of a bobolink swaying and singing on a spray of goldenrod. Beneath it was printed in black letters, MARY POWER."[18] The girls order cards at the town print shop, where letterpress messages would have

Figure 1-2   Scrapbook page circa 1880s, showing a mix of
calling cards and trade cards in similar formats. (Collection
of Bert and Ellen Denker.)

been applied to the chromolithographed cards sent to the town printer from
New York or Philadelphia.[19] In Wilder's narrative the cards themselves are valued and are traded enthusiastically: the more considerate girls give their name
cards to all their friends, while a notoriously selfish girl gives hers only to others who have one of these desirable cards to give her in turn, thus swelling her
own collection of colorful, pretty pictures.

Among children, the cards operated as markers within social interaction, a
sign of favor or inclusion. W. E. B. Du Bois, in *The Souls of Black Folk*, recalls a
late 1870s exchange of calling cards as a signal moment in which "a shadow
swept across" him, and he saw himself excluded by whites:

> I was a little thing. . . . In a wee wooden schoolhouse, something put it into the
> boys' and girls' heads to buy gorgeous visiting-cards—ten cents a package—and
> exchange. The exchange was merry, til one girl, a tall newcomer, refused my
> card—refused it peremptorily, with a glance. Then it dawned upon me that I was
> different from the others; or like, mayhap, in heart and life and longing, but shut
> out from their life by a vast veil.[20]

While the social calling cards represented the possibility of communicating exclusion through the refusal to give or receive them, the commercial cards would seem to have been available without such a social test. Even a child excluded from cliques based on race or based on having the money to buy calling cards might have access to the free trade cards. And yet the seemingly freer realm of commercial interchange had its own restrictions: a poor child's parents might not shop in places where trade cards were given out; the discourse of racist caricature on many trade cards would have conveyed its own message of exclusion from the consumer marketplace to a black child.[21]

Trade cards were desirable not merely as substitutes for calling cards but on their own terms, and advertisers recognized their attractions. An 1885 printing trade journal exulted optimistically that, "The number of people who save handsome advertising cards when they chance to get them is larger now than ever, and will increase with the growth of the population. No one is either so refined or so vulgar that he will not admire a pretty advertising card and save it. The ultimate destination of all cards is to swell some collection or to adorn some home, and they may be found in even the remotest parts of the land."[22] People saved the cards in part because color itself conferred value on them. The presence of color in an ad or art reproduction testified, as Neil Harris notes in his perceptive analysis of color and media, "to the greater wealth, ambition, or taste of its subjects or purchasers."[23]

Chromolithography was an additive process; multiple color impressions (at an additional cost per impression) achieved finer gradations in tonal range and produced a more lifelike and vivid image.[24] The asserted connection between the cost of the ad to the advertiser and its value to the recipient or reader was relatively straightforward: the greater expense to the advertiser resulted in a higher quality, more attractive card. Nathaniel Fowler's advertising manual warned not to stint on extra impressions: "Even the cheap label has a depth and warmth and character if given the advantage of an extra printing." Fowler compared extra printings to a "woman's underdress," advocating that advertisers give their work "all the underdress or underprinting that it deserves, and it stands out in the rounded perfection of healthful completeness."[25] Embedded in Fowler's slightly risqué analogy was a complex positioning of ads that rested on the fact that women had to be relatively well off to afford layers of underclothing while poorer women's clothing hung flat and presumably lacked character or distinction. Fowler implied that the ad itself was a luxury good—analogous to an expensive garment worn by the reader. The colorful, expensively produced ad in and of itself was a treat: both an invitation to step into a more "complete" and luxurious life by buying the product, and at the same time a token of that life. The more expensive an ad, the more it was worth to the consumer.

The overlap between the advertising and nonadvertising uses of stock cards rendered the advertising trade cards even more desirable than their physical qualities made them; distributors of chromolithographed novelties sold the same cards directly both to individuals and to advertisers. It was almost alchemy: the same cards that cost an individual from a penny to a dime each were given away for free when imprinted with an advertising slogan. It was a short step then, to the claim of one childhood collector who asserted that by urging his neighbors to buy a brand of coffee so he could collect its cards, he "made advertising pay"—suggesting that *he* was paid, that the cards were a kind of currency.[26]

The ubiquitous trade cards were a larger part of the child's visual experience than printed books,[27] and although chromolithographed advertising cards were plentiful, the scrapbook compilers treated them as though they were valuable. They saved them, preserved them within the often elaborate bindings of scrapbooks, and framed them on the page to maximize their appeal. By treating the cards as though they were valuable, collectors bestowed value on them.

By incorporating cards into what one commentator has called a salvage art, like some types of quilting, collectors fit cards into a paradigm of scarcity in which colorful printed material was valuable and must be saved, like other "scraps," a practice urged by housekeeping books earlier in the century.[28] In 1835, for example, Lydia Maria Child advocated "gathering up all the fragments, so that nothing is lost. . . . Nothing should be thrown away so long as it is possible to make any use of it, however trifling that use may be."[29] Defining the cards through the paradigm of scarcity that made a thrifty virtue of saving them, and then displaying them lavishly on the page, enhanced the sense of abundance the scrapbooks as a whole promoted. By amassing the cards on scrapbook pages, compilers created visual celebrations of plenty. They demonstrated that goods, or at least their surrogates—images of goods and cards given out on behalf of goods—were so plentiful that they were available for many sorts of treatment.

Coping with displays and representations of plenty was not a simple process, as Elaine Abelson's research on nineteenth-century shoplifters suggests. Playing with trade cards was one way to learn to "read" those images— to learn to desire but not be overwhelmed. Willa Cather's novel *My Ántonia* presents the contrasting situation of new immigrants who have not learned this language. Ten-year-old Jim Burden, recently arrived in Nebraska from Virginia, makes a picture book for his Bohemian immigrant neighbor Yulka Shimerda, Ántonia's younger sister. On the homemade scrapbook pages, Jim "grouped Sunday-School cards and advertising cards which I had brought from my 'old country.' "[30] The advertising cards are linked to the project of teaching Yulka to read and speak English.

Jim hears that the Shimerdas were pleased with their presents. But distributing this vision of American wealth, recirculating these gifts from the world of commerce, proves problematic: when the Shimerda family subsequently visits the Burdens, Mrs. Shimerda eyes the household goods enviously and demandingly, and is given a cooking pot; Ántonia tells Jim that his grandfather is rich and should help her family. While Mrs. Burden tells her grandson that it is seeing her children want for things that makes Mrs. Shimerda grasping, the juxtaposition of these scenes suggests that advertising cards effectively incite desire for a life filled with more goods. Jim has imagined himself generous in giving out pictures of wealth, tokens of plenty, as a sample of his heritage, but is offended when the Shimerdas understand them as direct representations rather than good wishes, and ask his family for the goods of which they have received pictorial samples. The Shimerdas are not literate in the reading process that Jim takes for granted.

Trade card scrapbooks tell us about how their compilers saw and categorized advertisements. Collecting ads and making scrapbooks flowed from other practices children were already encouraged in and were engaged in. What positions and roles did they construct for themselves in using the advertisements?

Collecting

Middle-class Victorian Americans were passionate collectors and classifiers.[31] Collecting was praised as good training in observation and in the valued skill of assigning and recognizing categories, of visually and tangibly taking possession of the world by naming and classifying it. Janet E. Ruutz-Rees's discussion in *Home Occupations* (1883) is typical of such praise. She speaks of an "inherent" "love of collecting."

> We recognize the budding taste in children who will carefully treasure up buttons, bits of tin-foil, beads, and glass to make collections; and again, in growing boys and girls, whose "fossils" or "shells" are objects of deep interest and amusement. . . . It is quite surprising to find how naturally interests spring up in connection with [collecting], so that in time the simple habit of taking care of things grows into one of classifying and arranging them.

Through such collections, "Valuable information is . . . acquired without any active effort."[32]

In Ruutz-Rees's use of the term, the "collecting" of postage stamps, of fossils, and of insect specimens are equivalent. In each, the collector learns habits of looking: "The eyes once opened to look for fossils, are awakened to a thousand facts hidden before. . . . A collector of fossils is like a person with second-sight. To him every particle of the earth's surface is fraught with meaning, and, his senses once fully awakened, new facts dawn upon him like a continuous revelation." Objects have a special status for the collector if similar specimens can be found in museums: "It seems to dignify the little bit of stone or crystal, and to make it a valuable acquisition at once."[33]

Within this 1880s rationale and understanding of collecting, seeking out and amassing trade cards was understood as sharpening the collector's eye to notice more trade cards and other forms of advertisement, to learn their language, and to learn about the world through them. Interests "naturally" sprung up in relation to them. Just as among stamp collectors, "no lad

who studies his stamp-book can be ignorant of changes of rule and nation-
alities," no girl who collected trade cards could remain ignorant of brand-
named goods, their claims, and the images and narratives associated with
each.[34]

Most trade card scrapbooks carry no indication of who made them;
although of those that do, a few were made by boys, trade card scrapbook
compiling was, by and large, associated with girls. While instructions for
scrapbook-making frequently appear in books of amusements for girls and in
some non-gender-specific books, they are absent from books for boys. And
among types of scrapbooks that might be made by male or female compilers,
the trade card scrapbook was specifically considered a female form: one 1883
discussion of scrapbooks of various types therefore switches from the generic
"he," used throughout the book, to "she" to discuss the trade card scrapbook
compiler.[35]

Boys engaged in a parallel pursuit in stamp collecting (see figure 1-3),
but they played within the discourse of commerce and nationalism and
learned its language, learning to put the rest of the world literally under their
thumbs. Nineteenth-century stamp collecting was praised for training boys
to participate in the marketplace of commerce, according to Steven M. Gelber's illuminating discussion of this practice: "By recapitulating many of the
fundamental structures and relationships of Gilded Age capitalism, stamp
collecting taught and reinforced the rules of the economic game." Stamp collectors learned to sort their stamps by country and by market criteria; sorting
according to aesthetic criteria was condemned. Although stamp collecting
would seem to be a logical extension of collecting the trade cards and other
chromos popular among girls and women, it is likely that it was because
stamp collecting was a masculine hobby that, acording to Gelber, "unlike the
commercially produced trade cards . . . stamps came to be regarded as
'important.'"[36] Inextricably bound with the gender of the children who took
up collecting stamps and trade cards was the relative value assigned to the
economic activities they recapitulated—participation in the marketplace as
traders and as sellers on the one hand, or as shoppers and tasteful arrangers
on the other. The result was that stamp collecting was valued while trade card
collecting was largely disregarded.

Trade card collectors engaged in the "classifying and arranging" of objects
that Ruutz-Rees values. The categories they sorted their cards into—in other
words, the classifications that made sense to them for these ads—had nothing
to do with the salability of cards. These categories were fortunately preserved
in scrapbooks, which tell us something of the lines along which trade card collectors trained their new "second sight," in Ruutz-Rees's terms, as they interacted with this advertising medium, learned to fantasize within it, and lived out
the cultural imperative to collect.

AN INNOCENT LOVE OF CARDS.   ARDENT PURSUIT OF STAMPS.

Figure 1-3   These panels from a drawing entitled "The Collecting Mania" classify trade card scrapbook making among other types of collecting and reiterate the idea that stamp collecting and trade card scrapbook making were a gendered pair. (*Lippincott's,* January 1882, p. 112.)

Trade Card Scrapbooks

The elastic term "scrapbook" described books holding materials very different from one another. Scrapbook making of all types was a popular hobby from the 1870s to 1890s, as evidenced in instructions in women's housekeeping manuals and in girls' books on pastimes and amusements. People can save "some little scrap in [the newspaper] which was of importance to them" in a scrapbook, along with poetry, anecdotes, and "valuable 'Hints' or 'Recipes,'" which will continue to have worth, a household hints book of the day suggests.[37] The ephemeral material in periodicals should be snagged and preserved, and the scrapbook was the place to do it: an 1892 *Harper's* column positioned scrapbooks as an intermediate category between magazines and books, a means of conferring literary immortality on printed ephemera. Complaining that "the magazine, in a generation that must run as it reads, takes the place of the book," the columnist asked, "Must we all go to making scrap-books in order to preserve the good things that fly on the leaves of the winged press?"[38] While scrapbooks could inculcate organization and foresight and lead to more thorough and complex ways to categorize knowledge, according to another writer, "It is a long way from such a scrapbook as this to the simple blank-book into which a child pastes the picture cards now thrust into her hands on every side."[39] And yet all types of scrapbooks train their makers, in Ruutz-Rees's terms, as the taste for collecting grows by what it feeds upon. E. W. Gurley's 1880 manual on making scrapbooks suggested, too, that under guidance, children "will soon show much skill and taste in arranging their little books, and also grow into a love for order and beauty."[40] The scrapbook not only trains its compiler, according to this writer, but also stands for the compiler: "We are all Scrap-books, and happy is he who has his Pages systematized, whose clippings have been culled from sources of truth and purity and who has them firmly Pasted into his Book."[41]

American scrapbooks of this period are of four general types: young children's picture scrapbooks, often with cloth pages; compilations of favorite bits of poetry and newspaper clippings; more personal memorabilia books; and trade card scrapbooks. However, the term "scrapbook" was used interchangeably for all types, despite their differences from one another, and blank scrapbooks, often with covers embossed or stamped "scrapbook," "scraps," or "album," were put to use as trade card scrapbooks, as well as for other types[42] (see figure 1-4).

Trade card scrapbooks primarily reflect the interest of the compilers in amassing and arranging several kinds of chromolithed materials. The cards' accessibility for playful arranging and pattern making from them is highlighted by contrasting them with European albums issued for collecting sets of trade cards, in which such play was rigidly structured. German and French albums made specifically for collecting series of Liebig's Extract of Beef cards, for example, had printed borders and precut slots for inserting cards. While it was possible to scramble a set or series, the incentive to do so was small, and the compilers of European books seem to have kept the series together by following the numbered order of scenes from *La Traviata* or the stories of the Round Table.[43] American manufacturers who gave out cards as premiums used similar strategies, using the consumer's desire to complete a set as an incentive to buy products in which the card was included. The account of one non-scrapbook-making childhood collector of trade cards mentioned earlier suggests that such collecting patterns were exactly what advertisers had in mind. Chalmers Lowell Pancoast reports that "as a school-boy I canvassed back doors urging housewives to buy Arbuckle's Coffee so that I could collect the travel and history cards. . . . I had the first complete set of Arbuckle's Coffee cards and as a collector of these advertisements became the envy of the schoolyard."[44]

Although sets were collected, they appear less often in the American books. Instead, as U.S. compilers amassed and arranged the chromolithographed materials, their scrapbooks reveal a variety of interactions with the cards, not always congruent with the advertisers' ostensible purpose of having recipients think of their cards in relation to the advertiser and its name.[45] Some compilers did associate advertiser and goods, and sorted cards according to product or manufacturer, regardless of the card's appearance. One compiler, for example, made a page of cards for products sold by Ayers—sarsaparilla, hair tonic, pills, and so on—grouping them despite the cards' different styles and formats. Another, following the lead of the manufacturer E. S. Wells—who made unlikely connections within a single card between the company's Rough on Rats poison and patent medicines such as Wells' Health Renewer and Mother Swan's Worm Syrup—grouped cards for all the Wells products on one spread.[46]

Another arrangement strategy, which organizes the cards by product cate-

Figure 1-4 Scrapbook cover, gilt embossed, circa 1880s. (Courtesy of The Winterthur Library: Joseph Downs Collection of Manuscripts and Printed Ephemera.)

gory, highlights the product and its uses. Compilers created galleries of comparative advertising by grouping together cards promoting thread or perfumes. This arrangement registers a high level of attention to the product's category and its uses as the compiler recreates a two-dimensional display counter of related goods.

Stock cards bearing the same picture of flowers or ducks might be used by many different advertisers and for calling cards as well. So some collectors grouped cards that were identical in design and differed only in their messages for different retailers, manufacturers, or people, or even with the message space left blank, as in figure 1-2. This juxtaposition of chromolithographed blanks and advertising cards might also call attention to the value of the blank card to the compiler, and to the coup of having gotten it free from the advertiser.

Similarly, other scrapbook arrangements group cards according to some element of likeness, crossing lines of product, manufacturer, and designer to group pictures of animals, of travelers, of children at the seashore, or of flowers, regardless of what the cards advertise (see figures 1-5 and 1-6). Such groupings override categories of products, as well as social categories: trade cards are pasted

Figure 1-5   Scrapbook page with pictures of children grouped around a portrait of Friedrich Froebel, originator of kindergartens. (Collection of Bert and Ellen Denker.)

next to calling cards, reward-of-merit cards (given out as prizes in school), Sunday school verse cards, holiday greeting cards, and the decorative die-cut embossed paper "scrap" sold for decorating goods and for making valentines.

Such arrangements seem to highlight the appearance of the cards and to ignore the product, but their reliance on the overlap between calling cards, blanks, and trade cards gives the trade card a quasi-familiar and even quasi-familial status. The advertiser becomes like a social caller, the friend or acquaintance who leaves a card on the silver tray. The format of trade cards, and the more overt encouragement later to play with ads positioned children's and other readers' interaction with the advertiser as privileged and social. The advertiser was thereby entitled to admission into the family circle on the same basis as a friend or family member.

Cards may embody projections: saving an advertising card of a sewing machine in its idealized setting expresses a compiler's wish for a happy and comfortable home, prosperous enough to afford a sewing machine and grant

Figure 1-6 Scrapbook page of trade cards and embossed scrap on nautical themes. (Collection of Bert and Ellen Denker.)

her a life free from tedious handsewing. Surrounding it with calling cards, as one compiler does, suggests that the idealized home includes a particular social network.[47] All the scrapbooks are in this sense wish books, but what's wished for may not be the merchandise, but the particular relationship to the world expressed through the cards. That relationship is nonetheless mediated by the advertising image. In fact, even as this compiler's calling cards speak of a specific and known set of social contacts, she in effect recreates the theme of another company's ads, which emphasized that their sewing machine was not an alien mechanical presence, but an integral part of domestic life and the home circle.[48]

Other groupings make elaborate, even quilt-like patterns, using the colors, intensity, or design of the cards as compositional elements (see figures 1-7 and 1-8). The popular quilt-making form of the 1880s and 1890s—the crazy quilt—emphasized richness of materials and layers of decorative elaboration, such as fancy stitching on top of velvet and silk cloth scraps. The designation *crazy quilt* meant that the work's focus was global, rather than the emphasis on repeated patterns of the block quilt.[49] The term also suggests something of the

Figure 1-7   Scrapbook page with trade cards arranged by
shape and color. (Collection of Bert and Ellen Denker.)

aesthetic of the chaotic that crazy quilts reveled in. In making patterns with the
cards, compilers demonstrated and developed their ability to make fine distinctions in shades and combinations of color. This ability was of heightened
value in dressmaking and home decoration, as the ever-finer subdivision of the
palette concocted by the growing dye industry brought more elaborate standards of color use to sewing. A middle-class woman might often be preoccupied with matching colors for sewing in the 1880s and 1890s; a feminist could
complain even in the 1870s that for women "as much time is devoted to matching a peculiar shade of olive-green as would go to the writing of an epic upon
Jerusalem."[50]

Duplicate cards were particularly useful in pattern making, and scrapbook
compilers often seized on the availability of multiple copies of identical cards
as an opportunity for other sorts of play as well. In one scrapbook that contains
many duplicates, numerous figures have been cut free of their surrounding
cards to allow freer play, and yet the commercial message is left intact.[51] Cards
have been manipulated to extract new narratives from those supplied by the
cards. A two-panel card for Rising Sun Stove Polish supplies the motif on one

Figure 1-8 Two-page scrapbook spread showing symmetrical arrangement of trade cards. (Collection of Bert and Ellen Denker.)

page. In the first panel, an angry housewife chases two sneaky salesmen who are trying to sell her an inferior product. In the second, a black couple and child kneel, staring at their reflection in the brilliantly shining stove. The compiler has pasted this card in the center and cut up a second copy and redistributed its figures in new configurations. The black couple now kneels before a large yellow marigold, probably taken from a seed catalog. Cutouts of the sneaky salesmen flank the card. Below, the angry housewife now points to another cutout of a white woman holding a girl on her knee. Possessing identical copies of the same material allowed this compiler to move characters almost magically from their original relationships into new ones. They step from their old frames into the world and take on an independent existence that moves away from, but continues to refer to, the original advertising source. The promulgation of many copies of the same image, decried by adult critics, enabled a new kind of children's play: the child celebrates the specifically mass-produced character of the materials in play and learns to stamp them with individuality.

This compiler's rearrangement, which sets an angry figure to scolding a particular family configuration, hints that, for this child, the scrapbook could speak of family relationships—something like a photograph album. This same scrapbook, whose opening page is titled "The Day of an Infant" and which begins with a series of cards following a toddler through a day of activities, later features a full page headed "Baby," with the word written in cutout letters (see figure 1-9). Instead of hand writing the word "Baby," this compiler stays within her strategy of using only chromolithed, mass-produced materials. The page insists on a relationship to both home and the commercial world. The compiler looks for her family in the marketplace of national advertising, and both finds it there and claims its place there.

The "Baby" page explodes with cutout chromolithed pictures of babies and young children. Its compiler uses the scrapbook as a proto-snapshot album, rearranging advertising images to refer to the compiler's own family, or to her own desires.[52] She surrounds herself with a new commercial family from these advertising pictures—pictures available to her manipulation. What field of choices did this compiler accept and reject in creating such a page? What cultural practices and assumptions did she draw on to do so? And what were the implications of these choices?

Having cut up trade cards, the compiler could have followed the advice of Lina and Adelia Beard's 1887 *American Girls' Handy Book*.[53] Complaining that scrapbooks of "pictured advertising cards" contain nothing but monotonous "row after row of cards," the Beards gave careful directions for cutting up advertising cards and reassembling parts to illustrate familiar nursery rhymes and stories. Their examples show figures cut entirely free from their trade cards, with brand names obliterated, removing them from their origins. But none of the scrapbooks surveyed did as they suggested. The intervention the Beards recommended would have restricted the marketplace attractions the cards offered their compilers. It implied that the pictures must be subsumed within a narrative over which adults had obvious authority and which was already considered appropriate for children.[54]

Figure 1-9   Trade cards, pictures, and letters cut from cards. (Courtesy of The Winterthur Library: Joseph Downs Collection of Manuscripts and Printed Ephemera.)

The Beards thus suggested that the topic of children's play be restricted to a properly noncommercial children's world. And yet scrapbook-making offered girls an area of apparent autonomy and acknowledged expertise in the arrangement of cards, as well as in a knowledge of modern products they had gained independently of parents or school and in which they might stand on an equal footing with adults. The new commercial discourse appeared to offer new freedoms, potentially lost in reinserting the cards into already acceptable topics of play. While trade cards were of course produced by adults, and many of the albums in which they were saved are elaborate enough to suggest that their purchase was bankrolled by adults, scrapbook-making seems to have offered children opportunities for play either outside of close supervision or in a private language, even as the cards themselves offered views of children in autonomous play.[55]

Examining late-twentieth-century children's relationship to advertising and commerce, Ellen Seiter finds commercial television directed to children similarly celebrating a community of peers. This celebration is "utopian, universally appealing to children in its subversion of parental values of discipline, seriousness, intellectual achievement, respect for authority, and complexity by celebrating rebellion, disruption, simplicity, freedom, and energy."[56] The iconography of some trade cards, too, may well have been treasured for its potential to be read as subverting the adult world. A card advertising the obviously adult purchase of a sewing machine, for example, shows a woman suffering the extreme embarrassment of having a poorly stitched dress seam rip, so that the skirt is torn from its bodice. Boys laugh and point, while her friend smugly tells her the name of the correct sewing machine to use, which will save her from such ridicule. The compilers who saved this card may well have savored its depiction of an adult authority figure in a humiliating situation. One compiler grouped it with another card on which adults are tripped up and embarrassed (figure 1-10). The fact that trade cards were considered of little value by adults seems to have made the trade card scrapbooks an area of relatively free play for children. They allowed expression mediated by the distant and apparently nonauthoritarian hand of commerce, rather than the close inspection of parents. Advertising entered the home as a seemingly well-behaved visitor, and then delighted the children by demonstrating new ways to make rude noises.

The creator of the "Baby" page, then, instead of setting the pictures of babies within household scenes or containing them within nursery rhyme narratives, preserved their ties to advertising. Compilers often echoed the imagery of the trade cards themselves in their imagery of plenty and abundance, amassing trade cards to create and reinforce such visions. Here the compiler presents virtual cornucopias of babies. This compiling, amassing vision pervades relationships with people as well as commodities.

Even as she preserves the pictures' ties to the commercial world, the compiler might still see them as even literally referring to her own family. Ruutz-Rees, in discussing scrapbooks, noted that compilers could choose pictures to identify with their own family and familiar surroundings: "The children who delight in scrap-books are not usually impressed by the style of the pictures. The charm for them lies in the details, in the figure of a woman that represents mamma, or of the kitten that is the exact likeness to the little beholder of the nursery pet."[57]

The photograph-like Lydia Pinkham card in figure 1-1 demonstrates one way of playing to this crossover between the intimacy of owning a photo of someone and seeing the photo's subject as an acquaintance.[58] Compilers similarly learned to identify their family with characters pictured in the trade cards and to use the advertising images for self-expression. The scenario in figure

Figure 1-10 Scrapbook page with two cards of adults in embarrassing positions. (Courtesy of The Winterthur Library: Joseph Downs Collection of Manuscripts and Printed Ephemera.)

1-11, for example, not only positions scrapbook making as an approved activity but also sets the young scrapbook maker in a central position, crowding out her more passive brother. The falling glue pot introduces a comic note of chaotic subversion, while a child old enough to make a scrapbook might also enjoy support for blaming the younger sibling in the picture who has just knocked it over.

Figure 1-11   Chase's Liquid Glue featured scrapbook making on its trade card both as a suggested use for the product and as a flattering portrayal of the scrapbook maker. (Courtesy of Russell Mascieri, Trade Card Collectors Association, Marlton, N.J.)

The choices its compiler made in constructing the "Baby" page illuminate the choices of other scrapbook compilers. We can see in her work ways in which advertising cards and later magazine ads might seem to their collectors, arrangers, and readers to speak for their desires or interests. This dynamic plays out more fully in one unusually complex scrapbook.

Trade Card Scrapbook as Personal Record

In the densest of the trade card scrapbooks reviewed for this study, one compiler interwove advertising cards with handwritten commentary that refers to the cards, to her own life, and to her social aspirations. Because the inside front cover has a handwritten card bearing the name Anna Skinner, I use that name for the compiler.[59] In her book, Skinner conducts a complex conversation with advertising.

For Anna Skinner, advertising cards and their pictures spoke eloquently to her wishes and experiences. Sometimes she amplified their conversation with her own notations, including religious verse, and elsewhere she illustrated and elaborated upon items of special personal meaning with other printed material. Page 14 of her book, for example (figure 1-12), has Lord & Taylor cards at each corner of the page, each different in subject and format. Below the card at top right, which bears a basket of forget-me-nots, Anna Skinner wrote, "'This flower is a sign to awaken thought, in friends who are *far away*.'" Skinner projected her wishes onto the card more articulately than the scrapbook compilers who pasted pictures of flowers next to pictures of groups of happy children, or calling cards around a sewing machine.

This page is thematically unified with cards that suggest foreignness or the exotic. They include: a small fan in the shape of a Japanese lantern and a card in the shape of a fan (advertising Mrs. Wheeler's Nursing Syrup)[60]; on the facing page, cards depicting sailors at sea using J & P Coats thread to tow Cleopatra's Needle as though it were a barge; a camel rider advertising Colburn's Mustard; the Pharos Watch Tower, from a series of cards on the Greatest Wonders of the World; and some cherubs "placing Cleopatra's Needle in position" with Merrick's thread (see figure 1-13). The world of commerce, then, speaks of travel and distance; grouping its images creates an accessible empire, domesticated through its connections to thread and mustard. A handwritten note below the Colburn's Mustard card links these cards to the Lord & Taylor forget-me-nots on the facing page: "'Forget me not, forget me *never* / 'Till yonder sun shall set forever.'"[61] Though handwritten, this message, too, is prefabricated, drawn from the conventional autograph book arsenal. The cards' allusions to exotic places are joined to specifically American references with another Lord & Taylor card in figure 1-12, this one a large die-cut Liberty Bell. To the bell's left is a cutout rooster, next to which Anna Skinner has written "'*Yankee doodle doo*' (*Chanticleer*)," adapting the conventional rendering of a rooster's crow into a nationalist one. Another organizational framework appears in her symmetrical placement of cards for three different thread companies, again referring to international landmarks and exoticism. Her arrangement of cards showing thread performing heroic feats resonates with the idea that the domestic spreads out to a larger world, even the unknown world: all the world's goods and clothing are hanging by a thread. At the same time, the cards refer to her own social life of "friends who are far away."

It is impossible to know what Anna Skinner meant when she pasted down her cards and wrote her notes. For example, the significance of the Lord & Tay-

Figure 1-12   Page 14 of Anna Skinner's scrapbook. Caption under the top left card: "This flower is a sign to awaken thought, in friends who are *far away*." (Courtesy of The Winterthur Library: Joseph Downs Collection of Manuscripts and Printed Ephemera.)

lor cards for her could lie in her associations with the store, but it could equally lie in the nineteenth-century coded language of flowers, in which forget-me-nots mean remembrance or true love; other flowers on the page offer similar possibilities for interpretation.[62] The flower on the card links itself as well to the "forget me not, forget me never" note on the page facing it: the two-page spread becomes a kind of totemic message transmitter, expressing her wish to be thought of or loved by someone traveling far away.

Figure 1-13 Page 15 of Anna Skinner's scrapbook. (Courtesy of The Winterthur Library: Joseph Downs Collection of Manuscripts and Printed Ephemera.)

Or she may have chosen the Lord & Taylor cards because they contained the word "Taylor": page 10 in her scrapbook includes in the top right corner a comparatively somber card displaying the cane-backed Taylor Chair. The Taylor Chair card is only half pasted down, and a small maple leaf has been tucked under it. Anna Skinner has written on the back of this card, "Christmas is for 'giving' and 'forgiving.'" This page appears to commemorate a trip; it contains cards for steamboats (one between New York and Boston, one unspecified), a card for the depot in Boston where the steamers arrive, a picture of a church with a full moon above, on which she has written "'The moon rose over the city, / Behind the dark church Tower,'" a New Year's card, a picture cut from a paper or magazine imprinted "Independence Hall, 1776," a card for mince meat showing a child happily eating mince pie, and three floral cards. Perhaps the assemblage commemorates a New Year's trip to Boston, on which mince pie was eaten; perhaps she visited the Taylors, whom she forgave for something, or who forgave her for something. The presence of Independence Hall in what seems to be a Boston assemblage could indicate Anna Skinner's geographic ignorance; it could be her way of commemorating a trip taken independently; it could allude to someone she met who was connected to Philadelphia.

The possible meanings of Anna Skinner's juxtapositions proliferate, and it is impossible to be certain about any interpretation. An underlying theme of her book is that more is more: she indefatigably layers her materials and writings, not so that any obscure others, but so that they embellish and frame one another, as crazy-quilt makers layered silks, velvets, and embroidery. Anna Skinner decorated the blank spaces between larger cards with deliberately selected bits of advertising matter, adorning even the gutter margins of her book with long narrow ads. And her assemblage was originally a source of more than visual pleasure: it included numerous perfumed cards scented with Hoyt's German Cologne and Austen's Forest Flower Cologne.

Using such adamantly public items as advertising cards, religious verse, autograph book inscriptions, and what appear to be other quoted materials, Anna Skinner created a private world, in which she fantasized through the language of ads as well as the language of flowers. Perhaps these public materials seemed to her so self-evidently to carry her sentiments, to be so articulate "a sign to friends who are far away," that she didn't need to explain herself more fully. And perhaps their value for her was that they held and carried her thoughts without overtly revealing them to those from whom she wished to conceal them—parents or prying siblings.[63]

Scrapbooks and Gender Training

The small number of women who ventured into stamp collecting in this period were often condemned for drawing on the same gender-linked skills and approaches that girls and women employed in scrapbook making: arranging items with reference to their aesthetic properties, or to make "novel and artistic designs."[64] But girls were sometimes recognized and actively solicited for scrapbook making, as, for example, by Protestant missionaries to India, who asked Sunday school girls to make "picture books . . . of advertising cards" for mission work.[65] Children were rewarded as well with a sense of competence in the growing national discourse of advertising.

An earlier nineteenth-century tradition had positioned advertisement as public amusement, celebrating the pleasures of exaggeration and humbug via medicine shows and extravagant claims. With the scrapbooks, we see advertising's offer of pleasure and amusement in itself, as well as in the advertised products, shift to a more private, interior ground. The cards were attractive: the girl was invited by the commercial discourse to come play with it. The invitation was sometimes a literal one, issued by the advertiser—the Lydia Pinkham portrait card bears the line "Put this in your album"—or the album itself was given as a premium to promote scrapbook making.[66] Girls learned to fantasize within the images of consumption provided, and they used the discourse of advertising to articulate and comment on their own fantasies.

The female compiler brought her aesthetic skills as an arranger of materials and as a decorator—skills produced by her gender-specific training—to bear on the arrangement of the cards on the page. She learned to create attractive and individual displays from mass-produced articles—applying "taste" to make something unique from goods everyone else also owned. Trade cards and scrapbooks appeared to expand the opportunities for play: the materials the compiler cut, arranged, and pasted were not precious, or even usable for something else, as scraps of cloth might be, but were obtained for free. She could cut them up and cut figures or scenes out of them without seeking adult permission, and without bowing to adult opinions as to what kind of play with them would be appropriate. The commercial world thus seemed to offer expanded opportunities and greater freedoms. For boys, too, cards might offer new opportunities—but these are more often noted as entrepreneurial springboards. Printing calling cards on a small hand press is mentioned, for example, as a childhood move into the world of publishing for Cyrus Curtis, later the publisher of the *Ladies' Home Journal*.[67] Such small presses could also add retailers' letterpress messages to stock trade cards.

And as we have seen, children seem to have resisted suggestions that they play with the cards in a child-appropriate way by using them to illustrate nursery rhymes, thus severing them from their commercial association. Cards offered girls a representation of both their own and their mothers' lives. They isolated and apotheosized the goods of daily life, frequently enlarging them to gigantic scale, as in trade cards for various brands of thread, in which a spool of thread looms as large in the card as it might in the life of a girl required to sew. (The cards that show the thread in this psychologically appropriate enlarged state often turn the spool into either a plaything or a vehicle of independence, as in figures 1-5, 1-9, and 1-13). In many trade cards issued by manufacturers, representations of the world of the home were central.

Trade cards made far more imaginative and elaborate use of pictorial elements than did periodical advertisements of the period, often wielding complex imagery with subtlety and sophistication.[68] Analyzing the cards of one manufacturer, the New Home Sewing Machine company, Maxine Carol Friedman finds that

> themes of home and home life, women's place in the home, even the courtship which would eventually bring a woman to reside in her separate sphere were all incorporated into the visual imagery of the cards of the New Home company. The message was . . . contained within the images, readily available to those conversant with these cultural codes, and would have been unmistakable to the female audience to whom the cards appealed.[69]

These trade cards were often sophisticated in their references to family ties and in their elaboration of scenarios of the social consequences of the use or nonuse of a product.[70] The scrapbook compilers' further embedding of the cards within their own fantasy lives both prefigured and oriented their response to later advertising strategies that more explicitly pointed to the social and emotional consequences of buying or not buying. Compilers primed their own ability to respond to advertising.

These cards set their dramatic or humorous scenarios amid the world of crying babies, of shopping well or badly, of problems caused by unfinished housework, or of work done quickly and easily with the advertised product, with happy consequences. Like the children who saved cards that showed adults in embarrassing positions, some compilers used the exaggerated claims of advertisers to comment on domestic life and work. The scrapbook page in figure 1-14, for example, has three representations of domestic life. In the card at the top, for Diamond Dyes, Mrs. Jones visits Mrs. Brown in her kitchen or laundry room; below it is "Baking Day," where three well-dressed women harmoniously perform baking tasks together while the Conqueror Wringer, which the card advertises, sits out of the way. Below these two calm kitchen scenes, however, is a scene of domestic uproar, in which an entire family throws the kitchen into chaos by chasing rats. And though "all this trouble might have been avoided by the use of one fifteen cent box of 'Rough on Rats,'" the picture of the energetic family and pets knocking over the baby in its high chair and upsetting the kitchen table in pursuit of the rat seems far more compelling. So, read from top to bottom, scenes of genteel kitchen industry are finally burlesqued and turned on their heads. As the compiler has arranged the cards, the realism of a household's dirty secrets, presented as low comedy, undercuts the pretensions of a kitchen presentable enough in which to entertain callers.[71] Here, too, advertising becomes a realm of licensed comment.

For girls, as subjects constituted within gender discourse, the space created by commercial discourse was attractive and perhaps less restrictive and less closely monitored than other activities. It gave them new playthings and new funds of knowledge and expertise, which carried a special status: even missionaries relied on girls' special knowledge of trade cards when calling them together to make scrapbooks. For middle-class women, too, the new institutions and cultural practices of shopping that became central to popular magazines in the 1890s appeared progressive and offered new freedoms.

Palaces of Consumption

Scrapbooks replicated features of the department store, embodying the department store style that Rosalind Williams has described as "chaotic-exotic."[72] In the 1880s, shopping played an increasing part in middle-class urban women's lives. Changing sale and display practices created worlds of visual and tactile luxury, both in the new department stores, where dry goods and other items were extravagantly displayed for sale, and in the grocery shops, where packag-

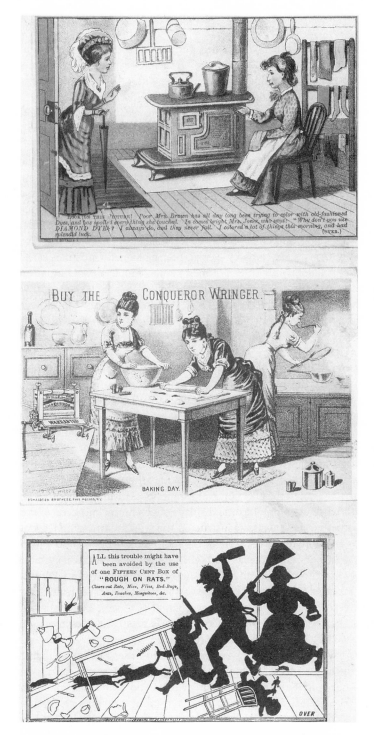

Figure 1-14 Representations of domestic life on one scrapbook page move from the genteel to the chaotic. (Courtesy of The Winterthur Library: Joseph Downs Collection of Manuscripts and Printed Ephemera.)

ing, distinctive and attractive labeling, and the prior training of consumers in the significance of brand names played an increasing role in sales. The urban woman's day increasingly integrated shopping with other activities such as visiting friends, dining, and being entertained.[73]

An early trade card scrapbook belonging to a girl from a wealthy Philadelphia Quaker family conveys the centrality of the retail world and shows the girl reconstructing her relationship to it.[74] Inscribed to Emily L. Blackburn, January 1st, 1874, when Emily would have been fifteen, the book mingles the personal, the sentimental, the social, and the upper-class retailer throughout, beginning with a lock of hair pressed in front of the first page of cards, all for Wanamaker's Ladies and Gents' Dining Rooms. Other cards from carriage-trade retailers follow—for Wanamaker's store, fur merchants, a children's clothing store, a "fine tailor," and an expensive candy maker.

Blackburn grouped cards by design, placing signed Christmas cards next to cards for retailers on the basis of their similar floral motifs. But criteria for acceptance into her book was also the cardgiver's place in her social world, since it was through that place that she got the card. The social rounds recorded in her scrapbook arrangements lead her to stylish shoemakers, past Christmas card senders, and through Strawbridge and Clothier and Wanamaker's Ladies and Gents' Dining Rooms.

Scrapbooks brought the physical patterns of shopping, and especially of the department store, into the home. In the scrapbooks were mingled the social and commercial, visiting cards and trade cards, and religious messages—as the cathedral-like department store mingled sales and concerts, rooms of merchandise obviously for sale, and ladies' parlors in which merchandise was displayed more subtly.

Though not new, shopping was newly emphasized as a social occasion. Shoppers walked the aisles of the mixed layout, stunned by the display; and at home, girls replicated the effect in their scrapbooks, turning pages to enjoy the superabundance of color, of exotic displays of costumes, and even of the customs of other lands.[75] Girls reconstructed the store, rearranging the colorful cards into a metonymy of the store in the scrapbook.[76] Frequent depictions of cascades of flowers, cornucopias, and other images of abundance, admitted the sensuous pleasure of shopping in the new department stores. (Figure 1-15 shows a scrapbook page whose cascade arrangement emphasizes the connection between abundance and shopping.) Scrapbooks were like the mail-order catalog of the 1880s and 1890s in which densely massed images in two-dimensional space also served as surrogate for the department store's three-dimensional display. But scrapbooks went beyond mail-order catalogs in their recreation of the experience of shopping. After all, wouldn't seeing friends be part of shopping? So the inclusion of visiting cards. And didn't retailers emphasize seasonal holiday sales and promotions, reframing religious celebrations as shopping occasions? So the holiday greeting cards. And might not a purchase be a treat rewarding achievement or right action? So the reward of merit cards. And wasn't shopping for goods for the sacred precincts of home in the cathedral-like department stores almost a religious exercise? So the biblical verse.

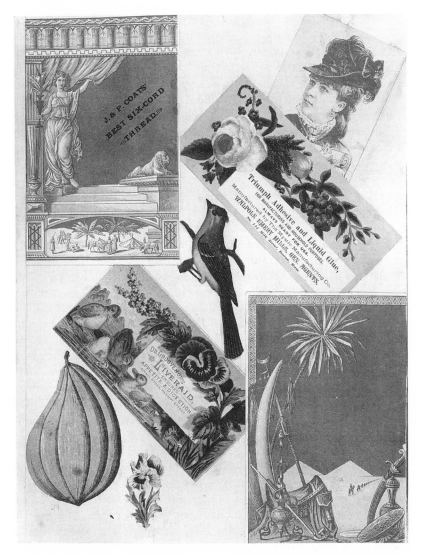

Figure 1-15 Scrapbook page cascade or cornucopia arrangement. (Courtesy of The Winterthur Library: Joseph Downs Collection of Manuscripts and Printed Ephemera.)

Both the scrapbook and the mail-order catalog were metonyms not only of the department store but of the magazine, with the catalog's loquacious copy as reading matter or news of merchandise, and the scrapbook's sharing with department store and magazine a mix of attention to the social, seasonal, religious, aesthetic, domestic, and exotic. Scrapbook makers naturalized not just their relation to national advertising, but the idea that ongoing, familiar ads had an expected place within any assortment of pleasures and entertainments. In actively creating their own version of such a miscellany, children invested themselves in it and took psychic ownership of it. The 1892 *Harper's* column that positioned scrapbooks as a way for compilers to intervene in the moving stream of periodicals by making their own compilations followed the same lines as it implied that readers might actively reproduce their own version of a magazine, thus investing themselves more deeply in it.[77]

No single card could be sufficient. Compilers amassed and even layered cards to magnify the sense of plethora and richness of color, and they formed compositions moving in different narrative and design directions. In the rare instances where something on a card offended, there was no need to exclude it. One compiler who included racy cigarette cards of stage performers covered their plunging necklines and bared legs with embossed paper scrap bouquets and even biblical sayings (a flowered "All things are possible to him that believeth, Mark 9:23" obscures an exposed leg).[78] There was room for everything if you just added more of it to fit it in. And, as in the Victorian parlor, anything could be covered to resemble something else.

The new department stores set everything out for display, inciting people to buy and even to shoplift. Shoppers experienced the constant buzz of merchandise, the erotics of display. A constant seduction: the seeming availability of goods, withdrawn at the cash box by the barrier of household means or the shopper's allowance. The shopper was invited into a store where everything was visible, a realm of visual opulence. She was invited to see this world as an extension of her home—not like a friend's home, but a place whose displayed goods she was invited to imagine in her own home, on her own children, dressing or adorning her own body. The shopper was invited in, but then thwarted: even if she could buy *some* of what was displayed, she could not take home everything she desired.

Department stores introduced a new set of interactions, Rosalind Williams notes. The customer could now freely browse and "indulge in dreams without being obligated to buy." The apparent freedom *not* to buy on any one visit to the store obscured the pressure exerted in pushing the visitor to want to buy *sometime*. In the department store, "consumers are an audience to be entertained by commodities. . . . Selling is mingled with amusement . . . [and] arousal of free-floating desire is as important as immediate purchase of particular items."[79] The department store as an institution promoted shopping as a practice, and shopping fed the desire to shop more.

Trade cards became both a tantalizing reminder of the visual abundance of the stores and a take-home sample of it—luxurious color, a surplus of information, and a surplus of desire.[80] Large department stores were still confined to big cities; a small-town child was more likely to keep a scrapbook than to visit Wanamaker's. Even more than mail-order catalogs, a scrapbook could offer participation from afar in the luxurious pleasure of shopping and thereby bring it near. A girl who preserved the experience or fantasy of shopping in her scrapbook kept the memory fresh: as one scrapbook-making instructor put it, a scrapbook is "not a grave in which you have buried all these good and beautiful thoughts, but a living treasure always open to your hand."[81] Collecting props as memory prompts was a late-Victorian preoccupation.[82] But while the scrapbook thus memorialized memorabilia and souvenirs of shopping, at the same time it preserved and replicated the department store experience of inadequacy: the single commodity (or advertising card) could only inadequately recreate the spectacular celebration of abundance in the department store display.[83] No single trade card was enough: more and more of them must be arranged on the page in a profusion of color and form. In the scrapbook as well as in the department store, internal consistency was not the point of the "hodgepodge of visual themes." Rather, the disparate and contradictory materials together served to distance the experience from the ordinary.[84]

The cards were gifts from the world of commerce, even trophies for shopping. That the cards were mass produced may have even made them more desirable: not only did their duplication offer new ways of playing, but also it gave the collector the pleasure of being like every other girl. As the national distribution of brand-named goods shaped a national vocabulary, trade cards contributed to a national children's culture: all over the country, children were pasting cards for Lautz's soap into their scrapbooks. A girl could compare her scrapbook with another's, imitate another child's work, or see what unique pattern a friend made of the same material.

Goods were, of course, produced in different regions of the country, in areas of diverse history, culture, and economy. The national distribution and promotion of goods, however, made consumption a unifying bond that overrode regionalism. The ad reader's awareness of a national culture of advertising defined any purchase or acquisition of a card as a participation in this national culture, or, alternatively, a nonpurchase as a failure or refusal to participate in it.[85] While scrapbooks from different regions include cards from regional retailers, the scrapbook itself, like the department store, was a unifying frame: trade card scrapbooks from different parts of the country are identifiable as a genre. Framing the cards within them became another way for the cards to move into the national culture, and in which the national culture was replicated in each home. Advertising thereby became a familiar part of a national culture of the home.

Not only were the cards mass-produced themselves, but they represented the still-novel mass production of packaged goods, each package a duplicate of the next. Collecting these duplicated cards brought into the home the novelty and glamour of this new commercial realm of mass-produced goods, in which everyone could possess an identical item. Just as for Ruutz-Rees the presence of the fossil in the museum "seems to dignify" the collector's "little bit of rock or crystal and make it a valuable acquisition,"[86] the trade card might have had greater value and dignity for the collector because the product it advertised, wrapped in the packaging pictured on the card, could be seen in a store. And like the fossil collector whose activity sharpened his or her eyes to geology, the collector's eye had been trained to look for the brand name.

Summary

Shopping changed; people took up new forms of consumption, and the new shopping was accompanied by new social forms. Even time changed: the new version of the calendar provided by Paris's Bon Marche, for example, no longer a record only of religious feasts and saints' days, now provided reminders of sales and "redefine[d] months of the year according to featured categories of fashion," as Rachel Bowlby has pointed out, while magazines presented seasonally oriented articles and ran seasonally oriented ads.[87]

The new patterns of consumption and the expectation that myriad pleasures would be available from shopping were inscribed in the topics of play taken up by children and adolescents as they became full-time consumers. And looking forward, we see that such play served to naturalize a new cultural form. This play adapted its forms from such familiar cultural practices as quiltmaking, memorabilia book making, the compiling of colorful scrapbooks by adults as picture books for small children, and the skillful cutting of cloth. The new advertising materials were grafted onto these existing forms to create a new form: the trade card scrapbook. This "naturalness" of the new social forms of shopping, of buying by brand name, and the proliferation of advertising, was actively produced through play within the new discourse of advertising for nationally distributed merchandise. It was produced through the active construction of imagination within advertising, as middle-class people in the 1880s and 1890s learned to live within the new discourse of consumption.

The promulgation of colorful trade cards and scrapbooks elicited consumer interaction with advertising. The mass produced, widely distributed cards became a medium with which children could enact the interplay between mass produced goods and the individual home that was becoming an increasing part of their daily lives. Cards also reinforced the sense that planning consumption or deciding what to buy could be yet another source of pleasure, and encouraged consumers to seek out, read, and collect more ads. Moreover, as a two-dimensional simulacrum of shopping that joined social, religious, commercial, and sometimes narrative pleasures, the scrapbook primed its compilers to interact with the magazine as another such two-dimensional form, and to see advertising as an indispensable part of it.

2

Training the Reader's Attention: Advertising Contests

From the advertisers' point of view, these competitions are most valuable. They make our young readers and their elders thoroughly familiar with the business announcements in the magazines, and cause these to be read and re-read with closest attention. The young people, too, learn much . . . and are entertained besides.

St. Nicholas, 1902[1]

As magazine advertising largely displaced trade cards as the preferred medium for advertising nationally distributed goods, and as publishers became economically dependent on advertising rather than on publication sales, magazines developed an institutional interest in training the attention of readers on advertising. Increased product sales following advertising would presumably demonstrate the worth of advertising in a magazine. But since advertisers' measures of how and where readers read ads were crude and scattershot, publishers attempted to demonstrate to advertisers that their ads were being read and to show that their magazines were a congenial environment for advertising.[2] In the 1890s magazines were already in the parlor. Their publishers now wished to demonstrate that they could make advertising, too, a welcome guest in the home.

To attract advertisers, ad-dependent magazines from the 1890s on frequently ran contests that invited readers—both children and adults—to play with the advertising. Publishers then publicized this advertising support work through ads and accounts in trade magazines directed to advertisers. While the attractiveness of trade cards invited readers to collect and arrange them, contests represented advertisers' more direct efforts to structure readers' imaginative interaction with advertising. Contests trained their audience as consumers and as magazine readers who would be attentive to advertising. Advertisers' own notions of how advertising influenced readers were still forming, and these contests tell us much about what advertisers sought to cultivate.

An ad trade journal announced that one 1904 contest, for example, offered money prizes "to readers selecting the best advertisements that appear in the *Booklovers.*' . . . It suggests a critical examination of the advertising pages, and secondly, it opens up a subject for intelligent discussion. Indirectly, such discussion *reacts to the advantage of all advertisers, giving them a correct estimate of their advertising copy*; for, after all, the public estimate is the only one worth anything to the advertiser."[3] What *Booklovers'* proposed was a far cry from the kind of market research that eventually developed and became widespread in the 1910s, which measured the effect of an ad on buying behavior. Initially, magazine publishers largely assumed that ads conforming to high aesthetic standards, the "best" ads, would naturally best appeal to the reader; they sought an aesthetic consensus on advertising.[4] This focus also helped to gain acceptance for the presence of advertisements in magazines: it asked readers to see them as aesthetic objects, existing within the same register as magazine stories.

Advertisers shared the belief that an ad had to be consciously noticed and carefully read. The notion that it need not be consciously seen, but rather uncritically absorbed, did not gain favor until the 1920s. It seemed logical to train readers as aesthetic critics of ads and to ask readers how advertising should look to be attractive to them. *Booklovers'* thereby invited readers to regard themselves as participants in advertising by including them among its judges and decisionmakers. As readers participated in advertising, they were encouraged to pay attention to advertisements and to "collect" the best ones, as they had trade cards, using the same scheme of attention as in Janet Ruutz-Rees's 1883 formulation of collecting discussed in chapter 1. They became more alert to advertising. The advertising "collector" was analogous to the fossil collector whose "eyes once opened to look for fossils, are awakened to a thousand facts hidden before."[5]

Readers collecting advertisements for a contest acquired a similarly finely tuned attentiveness to advertising and simultaneously learned new ways to play with them. Along the way, they learned to expect and even seek out ads in a magazine, and ads became a part of the magazine with which they felt particularly at home.

Children and the Psychology of Advertising

The new psychology of advertising developing at this time came not from Freud but grew out of William James's empiricist ideas, as explicated by psychology-of-advertising promoter Walter Dill Scott.[6] This approach sought to establish recollection of trade marks and to form habits of buying. Good advertising was a matter of intervening in the stream of consciousness with "suggestions," of shaping "perceptions," creating "associations," and making "impressions."[7] Scott's theories suggested that advertisements had to do more than grab attention, and that appeals to reason would not be sufficient to influence most readers to buy. Rather, Scott argued that habit, the fusing of pleasant associations with an ad, appeals to the senses, and appeals to action or "motor response," which would prime a reader to buy, explained the workings of successful ads.

Although children's activity as consumers was regarded as negligible, a body of early 1900s advertising trade writing dealt specifically with shaping the child's habits and impressions: "There is no period of life at which impressions can be so deeply made as in early youth, and as the youth of America are trained, so will the next generation of men and women be. Proper training ought, all educators agree, to consist in familiarizing young people with the conditions that will surround them in adult life."[8] Even as advertisers encouraged and endorsed a new, more malleable notion of "personality," of subjectivity or identity that shifted in response to surrounding needs and conditions, advertising culture laid claim to a more permanent character, which at least in childhood could receive a more permanent mark or impression.[9] So although Scott at one point endorsed a linear model of simple accretion of experience in forming a habit ("Every thought we think forms a pathway through our brains and makes it easier for every other similar thought"[10]), elsewhere he identified childhood as a special opportunity for creating associations and impressions:

> As a boy I associated certain names with certain articles of merchandise. I saw a particular soap advertised in various ways. Perhaps it was used in my home—I am not sure about that. The name and soap were so habitually associated in my mind as a boy that when I think of soap this particular soap is the kind I am most likely to think of even to the present time, although it has not been called to my mind so often in recent years as other kinds of soap. . . . The associations formed in youth are more effective than those formed in later years. The effectiveness is lasting and will still have influence as long as the person lives. Hence goods of a constant and recurring use might well be advertised in home or even children's papers, and the advertisements might be constructed that they would be appreciated by children.[11]

That is, within new models of fluid, shifting identity, nostalgia for the possibility of making a permanent mark, an assertion of stability, was grounded in the child, someone on whose mind the preference for a brand could be "impressed" or "engraved." Childhood became the repository of a commercial unconscious, where early habits left their untraceable mark on adult behavior. Scott believed in advertising "that appeals to the small boys and girls for firms that expect to be doing business say forty or fifty years afterwards, as it will take strong advertising in later years to make the small boy or girl choose other goods. Early impressions are the strongest and most lasting."[12]

As we have already seen, nineteenth-century children were avid readers and collectors of advertising. Writers of advertising advice gave two main reasons that ads for products intended for adults should be attractive to children. First, it was one way to get an advertisement into the home: if children collected attractive advertising paraphernalia and left it around, their mothers would have to look at it. Second, such advertising was an investment: children would be favorably disposed toward a product from an early age. Both of these ideas are central to the larger questions of how advertising's place in the home came to be naturalized and how consumption and advertising were constructed as arenas of pleasure and free play.

A body of early 1900s advertising-trade writing specifically dealt with the question of shaping the habits of the child to ensure later habits favorable to the advertiser. Girls in these writings were sometimes seen as newly minted consumers at marriage—envisioned as never having shopped for groceries before the honeymoon. In such scenarios, habits of consumption could be instilled entirely in the marketplace. Instead of choosing the brands they were accustomed to using at home, girls bought the brands of which they had most often heard in ads.

With the exception of the wealthiest fraction, girls of this period were drawn into housekeeping tasks at an early age and were surely familiar with the groceries in their family home. And yet the idea that family was not the chief source of shopping knowledge might have been attractive to girls. Just as collecting and arranging trade cards may have given children special knowledge and authority about products at least equal to that of adults, learning about products from ads could give young women an independent source of knowledge away from the family.

Writing sympathetically of advertising in 1904 in *The Atlantic Monthly*, its advertising manager MacGregor Jenkins pointed to the pleasures for the child of expertise obtained away from the home. He discussed the importance to the advertiser of inculcating children with brand preferences:

> Through the length and breadth of this great country thousands of men and women are daily, almost hourly, making their initial purchases of various wares.
>
> The comic papers have long made sport of the bride and her early experiments in marketing. But the establishment of each new home is a matter of importance to many advertisers; for once that their brand of soap or soup or silver polish be established in a household, the chances are it will remain the family standard for years to come. So the far-sighted advertiser begins to say "Pearline" to her in early infancy. Pearline follows her to school, thrusts itself upon her as she travels, and all unconsciously engraves itself upon her memory. The eventful day arrives—list in hand she sallies forth to her first day's shopping. Amid the confusion of new experiences she gloats over her ability to choose and purchase half-a-dozen common articles with the composure and accustomedness of a veteran. She orders Pears' Soap, White Label Soup, Pearline, Walter Baker's Cocoa, and Knox's Gelatine, because she knows and remembers the names, and does not realize that she has chosen in every instance an article made familiar to her, perhaps, by advertising only.[13]

The rhetoric of this scenario—in which, despite its feminine name, Pearline disconcertingly resembles a child molester following the girl to school and thrusting himself on her, thereby marking her for life—may have been distasteful to some advertising writers. By 1913, publishers wrote of trade names "gently impressed" upon their readers' memories, rather than "engraved" upon them. The child's mind, in this later metaphor, is malleable—subject to pressure of habit and ultimately molded by it—rather than already set, but capable of holding a single decisive engraved mark. In both versions, however, advertising to children is an excellent investment: it is the point at which an advertiser could gain a customer for life.

Regardless of which metaphor they preferred, publishers endorsed the idea that constantly repeating the product name to the child would secure its place in her store of childhood recollections and "unconscious" training. At marriage, the woman awakens as a customer from her childhood slumber.

> A young girl sees in the magazines on her mother's library table the advertisements of Gold Medal Flour. She has nothing to do with the buying of family supplies. Very likely she doesn't know what flour her mother buys. She doesn't even realize that she notices the advertising. The day comes when she marries, and soon after she gives her first order for groceries. She has to make a choice of flour, and automatically the name Gold Medal recommends itself to her as that of a good flour—one she has always known about.[14]

Jenkins calls buyers "members of this school, which the advertiser has been conducting with great expense and patience for many years" (see figure 2-1). The Pears', Baker's, and Knox's products he cites were all constant advertisers in the elite children's magazine *St. Nicholas*; from 1899, for over eighteen years, *St. Nicholas* conducted a monthly contest that encouraged readers to read and interact with advertising.

Advertising Appreciation

Like the *Booklovers'* contest, the *St. Nicholas* competitions sought to develop a critical and attentive eye; such conscious attention was believed to be better than the unconscious knowledge the girl in her mother's library had of ads: "For several years thousands of boys and girls have had their attention directed in such a manner as to compel close study of the advertising pages of *St. Nicholas* and *The Century* [published by the same company]. Every month tasks have been set them requiring them to examine and report upon the work of advertisers, or to make advertisements of their own."[15] *A Good Line of Advertising*, a booklet issued by *St. Nicholas* in 1904 to solicit advertising by explaining the ad contests, went on to point out the advantage to advertisers of its readers' training: "Readers of *St. Nicholas* have become the most intelligent constituency an advertiser can address, and a constituency that is of growing importance and value. The greatest advertisers in the country have recognized this, and have directly and indirectly lent their aid to this school of advertising, and of advertisement writers and artists."[16]

In the *St. Nicholas* contest, as in the *Booklovers'* contest, admiration and connoisseurship were seen as desirable responses to ads. As the editors tell their readers in the introduction to one contest in 1905, "it is necessary to educate in

The advertisement text reads:

> While looking through copies of *McClure's Magazine* an eight-year-old boy turned to his mother, and said:
>
> "Mamma, you know magazines are very useful. They tell you what you want, and where to get it."

THIS is an actual incident which shows strongly two things in regard to *McClure's Magazine* as an advertising medium. One is that *McClure's* is educating advertisement readers and, therefore, prospective buyers in the desirable homes of the country. The other is that it is reaching this constituency, which it has itself trained, with more desirable advertising than is carried by any similar publication to-day.

The boy is right. Magazines like *McClure's* do tell you what you want, and show you where you can get it.

McClure's Magazine is the marketplace of the world.

S. S. McCLURE COMPANY

Curtis P. Brady, Manager Advertising Department, New York.

Frederick C. Little, | Western Representatives,   Egerton Chichester,
Frederick E. M. Cole, | Marquette Bldg., Chicago. New England Representative, Globe Bldg., Boston.

Figure 2-1 A 1904 *McClure's* pitch to advertisers illustrates the scenario of a child educated by the ads in the library of his affluent family—here as part of a constituency the magazine has "trained." (*Profitable Advertising*, February 1904, p. 893)

the appreciation of good advertising not only those who will make a business of publicity but the whole public to whom their efforts are addressed."[17] The publishers were eager to show advertisers that *noncommercial* creativity would not be rewarded. Contest judges have "no use for the efforts of budding genius simply as such. . . . It is the object of these advertising competitions *to train young girls and boys so that they may become practical makers of advertisements*, if they choose, or at least may be appreciative of the work of others—an intelligent public to whom the advertiser may appeal with the certainty of a response."[18]

Adults, too, were encouraged to ally themselves with advertising by writing ads and reading them closely along the way. An 1897 competition in the mail-order monthly *Woman's World and Jenness Miller Monthly* similarly invited readers to take the role of ad writers: "We call your attention to our advertisements, with a view of developing the talent which we know exists among our readers. We shall award ten dollars to the writer of the best original advertisement of any article advertised."[19] Entrants were to "study" the ads "and try and improve upon them." *St. Nicholas* articulated the need to discipline readers to respond to ads, and it proposed the middle-class reader as the proper and desirable object of such training. While *St. Nicholas*'s contests prompted readers to try on the advertiser's role, and thus see themselves as having interests in common with advertisers, the magazine elsewhere directed advertisers to see *St. Nicholas* readers as young versions of themselves, and therefore worth reaching. One such message to advertisers assumed that the advertisers' own store of "engravings" and "impressions" was stocked by this genteel magazine:

LOOKING BACKWARD

You were brought up on *St. Nicholas* and remember still some of the stories, pictures, and possibly the advertisements which you saw there when you were young. *St. Nicholas* is the same magazine to-day. Don't you want to engrave upon the impressionable minds of the best boys and girls in the land your firm name, trade-mark, or commodity?[20]

And yet *St. Nicholas* was an elite magazine. E. B. White, claiming its readership was universal, actually revealed the opposite to be true when he wrote in 1934, "I suppose there exist a few adults who never even heard of the St. Nicholas League—people whose childhoods were spent on the other side of the tracks reading the *Youth's Companion*."[21] But since *St. Nicholas* claimed a circulation of between fifty and sixty thousand, and *Youth's Companion*'s has been estimated at half a million in this period, the other side of the tracks was clearly better populated.[22] Advertisers or ad managers might not have come from such well-off backgrounds, and other pitches proposed *St. Nicholas* as the advertiser's source of entree to such a household. Its "readers are constantly invited to close critical, intelligent scrutiny of its advertisements, which make them remembered. Examine any copy of *St. Nicholas*: what kind of family is it likely to enter? When you persuade the boys and girls that there is pleasure and profit in studying your advertisements, you surely reach the heart of the whole family."[23] An article on the competitions in an ad trade journal, after praising the nearly professional quality of the ads the children turned out, similarly concluded, "the greatest value, measured by a commercial standard, accrues to those advertisers who are shrewd enough to appreciate having an army of bright young people studying their announcements in the family circle, and so making their wares and trademarks household words in the best families everywhere."[24] Not only is the advertiser thus assured that the merchandise *St. Nicholas* offers it — the reader — is of high quality, but it is promised a seat at the table, entree into family conversation. The structure of the advertising competitions furthered this end.

"Everybody Try"

Working on advertising, playing at creating advertising, and playing with advertising slogans were all proposed as worthwhile family pastimes: "Remember that there is no age limit to these competitions, that you may have help, and that whole families may get together and combine on a clever idea."[25] Inserting ads in *St. Nicholas* "where they will become a part of these competitions" was the right strategy for advertisers if "you would like your products, your firm name, and trademark made the subject of household discussion."[26] Similarly, contest rules in a 1904 competition run by *Ladies' World* (discussed later in this chapter) encouraged contestants to make ad work a social project by offering a large prize to the winner and free subscriptions to anyone who collaborated with the winner.

Some *St. Nicholas* competitions extended the invitation to work together to the larger community of the school. Schools could earn reference books in a contest that asked advertising-related questions. The rules recommended even wider participation in the contest: "There is no restriction as to obtaining aid in answering the questions; on the contrary, each school is urged to secure the help of any one who is interested in their winning one of the prizes offered for the answers. Teachers and parents and grown up friends are asked to help their children to win the principal prize."[27] Even before adults were invited to enter *St. Nicholas*'s ad contests, it was reported that one family, at the father's suggestion, had gone over an issue and voted the ad contest the most interesting subject in the whole number. The family set to work with enthusiasm, first submitting the material to the mother for preliminary judging before the entire family's combined work was sent in.[28]

Other editorial matter helped to orient *St. Nicholas* readers toward contests and to identify with advertisers: at about the time the competitions began, contests became a frequent topic in *St. Nicholas* fiction. Similarly, both the contests and stories published in *St. Nicholas* showed making advertisements as an enjoyable social pursuit. In "The Corner Cupboard" by Margaret Johnson (1905), art students Kitty and Grace rent the small seaside cottage of the title to use for a studio and store, where they serve tea and sell fudge, decorative objects, and drawings to earn money for art school. But it goes badly; summer boarders are not buying. Their friend Billy has an inspiration: "Are we too proud to advertise?

Something unique and fresh in the way of posters is the very thing the Cupboard needs to make it go; something to catch the public eye—to fix the wandering fancy" of the summer boarder. They rewrite Samuel Woodsworth's sentimental favorite "The Old Oaken Bucket" as an ad. Billy begins:

> "How dear to this heart are the scenes of the seaside,
> When fond recollection presents them to view!—"
> "How gaily I wandered the sunny shore *be*'side," struck in Grace,—
> "Regardless of sunburn, or sand in my shoe!"
> "The pier and the plank and the little steam-ferry
> I've run for so often and always in vain
> The succulent crab and the wild huckleberry,
> And e'en the rude Cupboard that stood in the lane:
> The quaint little Cupboard, the trim Corner Cupboard
> The moss-covered Cupboard that stood in the lane!" They finished
>   all together in a jubilant chorus.[29]

The poem they are parodying is so familiar, and what should be done to it to make it an advertisement is so evident, that they perform the impossible task of thinking together, all reciting at once precisely the same words as they compose them. Producing advertising thus allows astonishingly social acts of creativity, an idealized closeness that bonds the group together as it taps broader cultural knowledge in service to an entrepreneurial goal.

The product of this brainstorming, put on posters, turns the tide and brings them business and finally the patronage of the governor. Advertising becomes the way for Kitty and Grace to help themselves, and yet to be rewarded for the character virtues of pluck and independence that inspire the governor's interest and patronage. Advertising is necessary to make the evidence of their industry and pluck visible. As in the advertising contests in the magazine, talent and noncommercial artwork aren't sufficient: advertising must be added. And yet advertising is presented as the most enjoyable part of running the business. It is, if anything, *more* creative and appealing than making the artistic knickknacks the girls sell.

The Two St. Nicholas Competitions

As we saw in chapter 1, advertising can appear as an antiauthoritarian realm of utopian community and play, especially to children. Late-twentieth-century ads dramatize pleasure in the realm of consumption and enjoyment of ads themselves is posed as an opposition to the dull realm of parental values such as discipline, order, and respect for rules. Consumption becomes the field for celebrating freedom and for disruption and antiauthoritarian rule breaking. Already in place in the early twentieth century, this opposition was embodied in the differences between *St. Nicholas*'s ad competition and its ongoing nonadvertising St. Nicholas League Competition. The latter competition ran in the body of the magazine and set readers under eighteen nonadvertising-related themes for poetry, essays, drawings, and photographs. The advertising competition ran in

the ad section of the magazine and set readers to various tasks. They included
creating ads on particular subjects or that used particular phrases, or in the form
of a dialogue between historical or fictional characters; they also included hunt-
ing through the magazine for ads in which certain pictures or phrases appeared
or completing a puzzle using ad phrases. The nonadvertising competition
required adult certification that a child had submitted an entry and that the
work was original.[30] The advertising contest eventually required no such war-
rant, did not bar copying, and invited entrants to obscure the traces of individual effort as they played with the already-existing texts of advertising. Advertisers, here represented by the "advertising editor" of *St. Nicholas*, invited readers to
the freest possible play—even transgressive play—with the advertising, Rules
against plagiarism would have been counterproductive since the advertising
materials were offered for playful appropriation by all. Ads thus moved into a
quasi-folk realm, consistent with the fostering of a new common language built
on nationwide familiarity with advertising characters and slogans. In another
sense, the companies extended their corporate largess in the form of the advertising figures and slogans: they allowed readers the treat of being praised and
rewarded for playing with private intellectual property as though it were public.

   Allowing collaborative group work on competitions not only made the ad
contest a freer area than the regular competition but also embedded products in
social situations and got people talking about them, something any individual
advertiser would like. The attraction to advertisers of having their products enter
family conversation became an attraction to children: advertising offered pleasures forbidden in the nonadvertising St. Nicholas League competition as it
commissioned children to quesiton adults on behalf of advertisers. Instructions
for one competition seemed to let readers in on the behind-the-scenes work of
the magazine. It set up the month's competition via a letter from *St. Nicholas's*
new advertising manager, Don M. Parker, making a suggestion to the judges:

> You have never given the boys and girls a single advertiser to work upon. Take
> just one and ask for new ideas. Take the first one in the book [a special position
> for which the advertiser usually paid a premium]—Swift & Company. Ask them
> to read Swift & Company's advertisements; tell them to talk with the man their
> mothers buy their meat from; ask their fathers how Swift & Company can do
> business upon a margin of three per cent. . . . There are at least a "Heinz" num-
> ber of varieties of reasons why Swift Premium Hams, Swift Premium Bacon, Swift
> Premium Lard or any other thing that is Swift & Company's which has been
> advertised in the St. Nicholas should continue to be advertised in the *St. Nicholas*
> Magazine.[31]

Like an earlier competition that told readers to "be on the watch . . . for inci-
dents showing the use or popularity of articles advertised in the magazine,"
Parker explicitly tells readers to seek out brand-name knowledge, thereby
infusing their conversation with their parents, neighbors, and local merchants
with it.[32] Moreover, his own writing is a model of lively, casual enfolding of
brand-name references ("There are at least a 'Heinz' number of varieties"). The
reader is told by example that this type of slang is acceptable, even within *St.*

Nicholas's standards of decorum. (Contestants had previously been warned to "avoid vulgar expressions and objectionable slang, so that your ads may appeal to people of good breeding.")[33] Parker's suggestion is followed by an exhortation further hinting that readers have a group interest, and even a family duty, in helping advertisers, while it advises bringing further group effort to bear on the competition entry: "Now go ahead, do it. Win the first prize by giving the brightest idea for the Swift & Company's advertising in *St. Nicholas*, and help them make their advertising even more interesting to the large *St. Nicholas* family. Get your best friend to help you. Everybody try."[34]

The ad contest emphasized the *process* of ad reading and writing rather than the product of such work. So while the main League competition printed winning work at length, the advertising contest page was taken up with detailed explanation of the rules for the next competition—and most often the prize-winning work had already vanished. While some ad competition tasks involved readers in apprentice-like efforts—asking them to create ads like those surrounding the competition page, and to thereby identify with the advertisement creator and to identify their interests with those of the advertiser—the physical product of that communion was accorded little significance.

Unlike the winners of the nonadvertising League competition, advertising competition winners rarely contributed visibly to the contents of the magazine. And even recognition for their work was ephemeral, stripped out of the magazine along with the rest of the advertising section when sets were bound for libraries or home preservation.[35] But by 1911, entrants in the advertising competition were being told that their work *as consumers* was helping to create the magazine:[36] "We are proud of our advertising department and want every one of you boys and girls to take more interest in what we are doing from month to month. The more interest you show in these pages the more able we are to secure for your reading and study interesting and instructive stories about all sorts of good things."[37] The advertising work readers produced served as evidence of their attentiveness as consumers, and thus enhanced the magazine. And yet the contests put less emphasis on the advertised product and more on the instructions for how to reenter the world of advertising and what to look for next. The ads were evidently regarded as so rhetorically powerful that study of them would produce the desired effect: that is, to leave "impressions" for later use. Moreover, the contests' requirement that readers respond in writing to the ads accorded with Walter Dill Scott's prescription for making ads memorable, which held that the reader's "intensity"of response to an ad would cause him or her to remember it:

> The intensity of the impression which an advertisement makes is dependent upon the response which it secures from the readers. The pedagogue would call this action the "motor response," even though it were nothing more than the writing of a postal card. Such action is vital in assisting the memory of the readers. An advertisement which secures a response sufficient to lead to the writing of a postal card has a chance of being remembered which is incomparably greater than that of other advertisements.[38]

This principle is at work in instructions for a 1911 "letter writing competition":

> The Judges want you to write a letter to a friend, real or imaginary, in which you describe the virtues of some article advertised in this number of *St. Nicholas*. . . .
>
> Don't begin the letter until you have sat and thought about your subject— go over all the advertisements carefully before you make up your mind which one to choose, and then write to your friend about it.
>
> You may tell how you used this article yourself; or any other facts that will show that *you understand the claims that the makers set forth*. . . .
>
> Your friends the judges get a little boastful sometimes about your abilities, so take care not to disappoint them. Make your manuscript just as real as you can, so that each will be a credit to you as a member of the great *St. Nicholas* family.[39]

The emphasis on eliciting social involvement with advertising materials and products is dramatically illustrated in two winning responses. The first letter, by John Ketcham, smoothly touted the superior ability of LePage's liquid glue over flour and water paste for holding together model airplanes, embedding the mention of the glue within the writer's and correspondent's mutual interest as hobbyists.[40] The letter of the second published winner, Elliot Weld Brown, written to his cousin, seems less precisely on point than Ketcham's. The letter's scenario, however, inserts advertising into the family far more firmly:

> Dear Freddie:
>
> Look on the back cover of the March number of *St. Nicholas* and see if you think the boy in the "Colgate's Ribbon Dental Cream" advertisement looks like any one you know! Mother says it's the image of *me* only I haven't got curly hair, thank goodness. Mother says she's going to take a picture of me in the same position, brushing my teeth, just to compare them, and *I* use "Colgate's Dental Cream" too, do you? And don't you love it? Mother says I never brushed my teeth so often or so long, as since we began using "C.D.C." It is such fun to squeeze the tube and let the cream run out so neatly onto the tooth-brush, don't you think so? And it leaves such a good taste in your mouth, yum, yum. Just as if you'd been eating candy. . . .
>
> I used to fill out the postal cards that came wrapped around the tubes with the names of my friends, and they received sample tubes of the cream. One day I wrote down, just for fun, "Miss Beauty Walter," the name of Uncle Frank's and Aunt Vic's little white dog, you know, that they always speak of as "your little cousin Beauty," just as if it were a real child.[41]

Colgate sends the dog a sample with a card addressed to her saying that Brown had sent her name. But now uncle and aunt use Colgate's, "so Colgate & Co. ought to forgive me for the joke I played on them, don't you think so?"

Evidently it did: the Colgate company seems to have seen advantages for itself in this technique of encouraging play with the product and its advertising, since it announced its own similar contest in the same issue (see figure 2-2). Brown's scenario demonstrates ways to play with and around the product, identifying many social uses as he makes it part of his conversation with his friend. Its ads are the occasion of comparison: he looks just like the boy in the

Write us an Advertisement
53 Prizes — Your "Ad." May Win One of Them

COLGATE & CO.'S advertising manager has been wondering whether he knows how to write the *best* kind of Ribbon Dental Cream advertisement for ST. NICHOLAS readers, and nobody can tell him this quite as well as the ST. NICHOLAS readers themselves. That is why this advertising contest has been decided upon and each one of you boys and girls is invited to compete.

There is nothing mysterious about this business of writing advertisements. The chances are that you'll find it much easier than your last school composition. Just imagine that you're writing a short letter to one of your schoolmates telling how important it is to take proper care of the teeth and how Ribbon Dental Cream is not only the best of cleansers but besides is so delicious in flavor that its use is a real treat.

Then take off the "Dear Agnes" or "Dear Bob" part of it and *probably* you have a good Colgate advertisement. We say "probably" because if your letter were just *sentences* about Ribbon Dental Cream without any real feeling of sincerity behind them, then Agnes or Bob would pay very little attention and it would *not* be a good advertisement.

An advertisement ought to *convince*, and you can't very well convince others until you have convinced yourself.

So before you write the "ad." be sure that *you* realize how important to *your* health and appearance clean teeth are. Read the Colgate "ads." in the back numbers of ST. NICHOLAS and in other magazines. Ask your teacher or your parents what they know or have heard of the great Dental Hygiene movement that is spreading throughout the country. *Try* Ribbon Dental Cream, if you are not already using it. You can get it at your druggist's, or if you send us 4 cents we will mail you a little trial tube, together with a Good Teeth-Good Health pledge card that has helped thousands of boys and girls in the daily care of their teeth.

We are giving you over a month to write the prize advertisement and our advice is *not* to write it until you really *believe* every word that you write. And remember, the *more* you believe it the *easier* it will be to write it and the *better* the advertisement.

RULES OF THE CONTEST

Any reader of ST. NICHOLAS under twenty years of age may enter.

Each contestant is limited to three advertisements, which must be on separate sheets of paper or cardboard, hand-written in ink, or type-written. Advertisements must not be more than 200 words.

Pictures, drawings, diagrams or photographs, illustrating your advertisements, are not required, but good ones (particularly photographs) may help you win a prize. These illustrations or "layouts" must be arranged to fit one full page in ST. NICHOLAS and should measure 5½ inches wide x 8 inches deep.

Advertisements may be sent folded, flat or rolled, but do not fold photographs or drawings.

To be considered they must reach us before August 5th.

Prize checks will be mailed the winners on Sept. 1st, and the awards will be announced on the back cover of the October ST. NICHOLAS.

Place sufficient postage on your envelope and address Colgate & Co., Advertising Contest, 199 Fulton St., New York City.

PRIZES

1 First Prize	$15.00
2 Second Prizes, each . . .	5.00
5 Third Prizes, each . . .	3.00
15 Fourth Prizes, each . . .	2.00
30 Fifth Prizes, each . . .	1.00

JUDGES

Mr. Sidney M. Colgate of Colgate & Co.

Mr. Don M. Parker, Advertising Manager of St. Nicholas

Mr. Francis A. Collins, author of "Model Aeroplanes"

COLGATE & CO., 199 FULTON ST., NEW YORK
Makers of the famous Cashmere Bouquet Soap.

Figure 2-2 Colgate's own advertising contest in *St. Nicholas* echoes the magazine's contest instructions. (*St. Nicholas*, July 1911, ad p. 13)

ad, and his mother even plans to take a picture of him modeling himself on that boy; the ad furnishes subject matter for a photographic tableau vivant. To properly step into the ad (as in carnival photos in which people stuck their heads into painted bodies), Elliot Brown must take up the proper product as well as the pose. While holding the toothpaste, he tells his friend about the incidental as well as the utilitarian pleasures of using it.

The letter features many ways to incorporate the product into play, social interaction, and even social commentary. As Brown acts as Walter Dill Scott's ideal ad reader, bringing his memory-aiding motor response into play by filling out the cards for samples, and thereby modeling this action for the *St. Nicholas* readers, he gives quasi-gifts with them. He sends one of these to a dog whose owners treat it too much like a person. Like the children who saved trade cards of adults in embarrassing positions, Brown finds license in advertising to mock adult authority: his knowledge of this realm lets him wield the power of a large company within his own family. Aunt and uncle respond to this act of social commentary with delight, showing the card addressed to their dog around to everyone who calls—they are perhaps amused that a big company, too, is willing to treat their dog as a person. And, in the Brown scenario, all of this feeds on itself, producing neither more social commentary, nor a change in how the uncle and aunt treat Beauty, but more consumers. Everyone is converted as the product and its advertising pass among them, and all are left with a surplus of good feeling. The pleasures of posing for pictures, of teasing older relatives, are all claimed in the name of advertising.

Advertising Games

As we saw with the trade cards, young readers needed relatively little encouragement to play with advertising. Novelty, widespread availability, and familiarity—the fact that the advertiser had already done the work of making a reference recognizable—seem to have been enough to make ads attractive material for play. Books of amusements often suggested ads as a source of parlor games for adults as well in this period. Several games made up an "advertisement social," for example, proposed as "a new and striking evening's entertainment" in a 1902 amusement book. The host collects ads, trims the words away, and numbers and displays them. When guests write the name of the article advertised on their sheets by number, "a merry and perplexing time will be the result. Although nearly all the pictures will appear familiar, yet few will remember what article they advertised. When the lists are completed a prize . . . should be given the most successful contestant."[42] Similar memory games were suggested in these and other books to reward knowledge of biblical verses. In the advertising memory tests, however, the players' inability to remember the material is hailed as a source of amusement, since the material is so ubiquitous that everyone expects that it can be easily remembered.

Ads are a cheap, available source of graphics in another game in the advertisement social, in which ads are cut up as jigsaw puzzles. The guests who put the ads back into their original form draw on their memory of the ad's appearance before it was cut up: the ad pictures have become so familiar that fragmenting them and making them whole again is a source of pleasure. In a third part of the social, ads are again sources of visual images, and the pleasure in the game again refers to familiarity with the ad, as players write copy for an ad picture chosen by the host. When the ads thus produced are read aloud, "the result will be much merriment." The laughter presumably rewards both outlandishness and the incongruity between the picture's original use and the new subjects chosen for it, a sense of incongruity that depends on knowledge of the advertising image and the feeling that the original, familiar version is the right one.

Because these games were played independently, without direct guidance or reward from advertisers, players were allowed the pleasure of mocking or pulling apart ads. Parody was a source of enjoyment. The *St. Nicholas* contests, by contrast, were run with a more firmly directive hand. Excessively free play with ads made the editors nervous at first, since the ads and contests sought to produce brand and product awareness in readers. They were to demonstrate to advertisers that ads in *St. Nicholas* successfully engraved or impressed names on readers and that readers were attentive pupils in "this school of advertising." The editor initially warned readers against playing with advertising material in a fashion that might indicate lapses of attention, or intent to parody:

> Try to remember that an advertisement is primarily intended to sell goods. It is amusing perhaps to receive suggestions that "Pears' Soap floats" that it may be "used as a laundry soap," but suggestions of that kind are not likely to win prizes. The St. Nicholas Leaguers have advised the cleansing of Mary's Little Lamb with "Pears' Soap," "Wool Soap," and "Sapolio"; they also hint that Sapolio [a scouring soap] is "good for soap bubbles" and for "cleaning silks" . . . and make other amusing illustrations of inattention.[43]

A later *St. Nicholas* contest, however, invited readers to the wilder, freer play of making comical collages of different pieces of ads (see figure 2-3.) This contest did not just license burlesque and parody, but capitalized on the status of the ad pages as something a parent might allow a child to cut up. And yet the humor here depended on familiarity with the proper form of the ad so that the reader could fully appreciate its incongruity with the new form. The *St. Nicholas* editors suggested a form of play with the collages that would restore the "educational" function: readers could display the collaged ads at a party while guests guessed where the different pieces came from.[44] Ads, and the magazines acting in their behalf, enforced both knowledge of product categories themselves and the sense that such categories were important. The "educated" consumer would realize that there were many kinds of soap, some appropriate for washing lambs, others for blowing bubbles, and still others for scrubbing floors, and that it was important to choose the correct kind.

Even when ad games encouraged playful parody, they both relied on and taught ad conventions and categories, as in a parlor game in which players wrote advertisement copy without naming the article advertised.[45] Like the *St. Nicholas* contests, these games encouraged readers to study ads. The players

Figure 2-3 Two winning entries in the Advertising Patchwork Competition, which called for "the most amusing or surprising combination of text and pictures from the advertising pages." (*St. Nicholas* August 1902, ad pp. 10 –11)

taught one another to notice clues to what was being advertised; they learned the developing conventions of advertising for different types of commodities; they practiced connecting products with what may have seemed an arbitrary set of attributes assigned to them. Even when games focused on the absurdity and arbitrariness of the newly emerging advertising conventions, the "merriment" that these games were said to produce rewarded players who recognized the slogans and characters of the ads, and who played within advertising conventions. The games' idealized result is embodied in the children of "The Corner Cupboard," so familiar with the relevant conventions that they can compose an ad simultaneously.

Advertising games were not merely suggested in books, they were actually played. For example, in 1903, the Young Folks Union of a Detroit church rejected other suggestions for an entertainment as "old, stupid, impractical or too much work. At last a wide-awake member proposed an 'Advertising Party.' The idea was novel and the tired minds seized upon" the suggestion for an evening of advertising tableaux vivants. After each "living picture" of a well-known advertisement was exhibited, the audience wrote down the name of the advertisement and any verse or motto that usually accompanied it. The report of this event in an advertising trade journal notes that:

> The Spotless Town tableaux were probably most effective, and their respective verses were the hardest test for the memories of the audience. The policeman,

clad in a suit gladly loaned by the police department of the city; the maid, with her pail; the doctor, just discovering the case, and the butcher, armed with his huge knife were each presented alone.

Then, later in the evening, the combinations of different inhabitants and all the inhabitants of Spotless Town were presented with great success.[46]

The church group finds this entertainment "the pleasantest of the year." The endlessly combinable images and characters of Sapolio scouring soap's Spotless Town ads invite people to play with them, and delight the audience with their versatility. (They were therefore more successful than the same entertainment's "'Good morning, have you used Pears' soap?' lady [who] drew the curtains back and smiled a two-minute smile at the audience.") Even the Detroit police department is pleased to join in and be identified with Spotless Town's admirable police force. The audience's inability to remember the exact verses perhaps allows them to believe that they don't pay attention to advertising, and, in playing the game, tantalizes them and demands they focus on the slogans.

Given that this entertainment was proposed by a "wide-awake" member of the group, it's not surprising that the evening's prizes frame attention to advertising in similar terms of intelligence and progressive thinking: the person with the largest number of correct guesses won a "cake of Sapolio, bearing the inscription, 'You take the cake for brightness.' The booby prize was a bottle of catchup awarded with the hope that its owner would catch up."[47] The church entertainment context, in which tableaux of biblical scenes or proverbs would have been more traditional, brought advertisements and advertising figures within an approved circle. The entertainment value of advertisements and advertising figures was no longer of a piece with medicine shows and the attraction of exaggerated claims; rather, advertising figures were a charmingly accessible and manipulable part of the family, encountered in respectable surroundings like magazines, or in the church or family parlor. One *St. Nicholas* contest even invited its readers to create a shadow family of sorts: the Uptodates, who had experiences "with modern advertised articles."[48] Despite their place in the home, these ad games were charged with a sense of adventurousness, of being modern and wide-awake, possibly even transgressive. Participating in advertising was thus both safely respectable and daring.

Other Contests

Other magazines recognized the value of contests for both involving readers in ads and demonstrating the readers' involvement to advertisers. *Good Housekeeping*, for example, rewarded attentiveness to ads in its 1903–1904 contests with rebuses and puzzles in which entrants identified a brand name from pictures (figure 2-4). (The simplest was a picture of a Quaker's hat with a sprig of oats in the brim, standing for Quaker Oats.) Once they had attuned their thoughts to the brand name references, entrants were to write testimonials on their presumably positive experiences with the products mentioned—another

Figure 2-4 Advertising contest. *Answers*: 1. Gold Dust; 2. Columbia Yarns; 3. X-Radium Heater; 4. Silver Salt; 5. Rogers Floorstain. (*Good Housekeeping*, April 1904, p. 438)

sort of ad writing.[49] *Good Housekeeping*'s consumer testing department, with its famous seal of approval, similarly reported only on products' laudable qualities. Its contest additionally directed attention to advertised goods by having winners select several dollars worth of merchandise advertised in the issue of the magazine in which the contest appeared. The hope of winning would naturally inspire interest in browsing through the ads to consider what choices to make if the prize came through. Entrants were directed to play imaginatively with the ads and to fantasize about having the products. The contest prizes were a license to consider accessible even merchandise that was outside readers' budgets, and to see the magazine as the proper forum for such fantasy.

Readers were invited into a more complex pattern of play when the inexpensive mail-order magazine *Ladies' World* offered what an ad trade journal termed a "decidedly novel" contest in 1904. Stories in this magazine posed authorship as an almost emblematic way for women in the home to earn money and linked such earning to spending money on goods and services to reduce household labor, as chapter 5 will discuss in more detail. The connection these *Ladies' World* stories made between authorship and purchasing power, writing and shopping, was strengthened in one of the magazine's strategies to attract advertisers: a contest in which readers wrote stories and poems incorporating phrases drawn from one issue's ads.

The contest not only ensured that readers would read the ads thoroughly while looking for phrases but also promoted the reader's more intimate relationship with the advertising characters and slogans. The contest rules invited readers to create an imaginative work from the advertising materials, as the scrapbook makers did with their trade cards; they were to fantasize through the materials of advertising. The model stories presented with the rules recommend by example that contestants see the figures in the ads as characters available for play and manipulation, that is, as characters who could be cut free from their advertising context and incorporated into other fantasies. While scrapbook compilers most often retained the link to the advertising, nothing required them to, and while the tableaux presented at the Young Folks Union advertising party retained the link voluntarily, they simply replicated ads. In contrast, rules for this contest required entrants to nod to the sponsor of their enjoyable play: they were to "give full credit, mentioning after each sentence or portion of a sentence, the advertisement from which the quote was taken."[50] They thus preserved the association between the phrase or character and its product while moving it into a new context.

Nineteenth-century readers already had considerable practice in granting to characters lives separate from the works in which they appear; they demonstrated this approach in, for example, the popular drawing in which characters from different Dickens novels surround Dickens's empty chair after his death. Books along the lines of Mary Cowden Clarke's popular 1850 *Girlhood of Shakespeare's Heroines* similarly freed characters from the boundaries of their plays, allowing Clarke to "endow them with rich lives of their own," as Nina Auerbach puts it.[51] The advertising trade press recognized how valuable advertisers found characters that could be taken up and brought into new contexts; they gloated over the free publicity supplied by such successes. As an editorial in *Profitable Advertising* exulted, "the value of a catchy phrase is even exceeded by the value of a catchy character, while a happy combination of phrase and character means assured success for the lucky advertisers."[52] The editorial pointed to the N. K. Fairbank Company's triumph with its Gold Dust Twins, used in its washing powder advertisements—caricatures of two young black boys slaving away at various household tasks, with the slogan, "Let the Gold Dust Twins do your work." *Profitable Advertising* noted, "These little darkies have the virtue of adaptability, the result being that they have been seized upon by cartoonists throughout the country, and during the last political campaign they have been conspicuously in evidence."[53] Such uses "serve to demonstrate the great value of adaptable characters to an advertiser. Every Gold Dust Twins cartoon published was a free advertisement for the Fairbank Company's washing powder, appearing in the character of 'pure reading matter,' and serving to make the twins and their product more conspicuous and famous."[54] *Profitable Advertising* emphasized the advertiser's coup in *creating* characters capable of being taken up in this way. The contests, on the other hand, worked the other side of the interaction between advertiser and reader: they prepared readers to *notice* such characters, to take them up, and to make them part of their thoughts.

Advertising characters were brought to narrative life in the sample stories offered as patterns for the *Ladies' World* contest. Given the necessity of incorporating product names and odd phrases, shopping turned out to be the ideal activity for these characters. In "'Be[ing] Sunny' (Force)," for example, Sunny Jim and Miss Cottolene, representing Force breakfast cereal and Cottolene shortening, respectively, are transported to a store and enjoy being surrounded by products:

> With "a smile on his face" (Sat. Evening Post) Jim took off his gloves, which were fastened in a queer way by "Peet's Invisible Eyes" (Peet Bros.), and drew out a handkerchief marked by a "Wasche Medallion" (Kaufman Medallion Co.), saying: "My motto is 'Be Sunny'" (Force).
>
> "You help yourself, Jim, I am tired," she said, sinking into a "Morris Chair" (Montgomery Ward & Co.) and drawing out some "Indian Bead Work" (Shell Novelty Co.). She watched him lazily for a while and then said, "Jim, what a terribly high collar you wear. I know it hurts you." "But that 'isn't true'" (Force), he replied. "'It hasn't scratched yet' (Bon Ami), and I like it, because by holding my head up 'it quiets the heart action'" (Pabst Extract).[55]

The writer here draws on the reader's familiarity with Force cereal's character Sunny Jim, always pictured with a long neck and a high collar. The language of this passage turns surreal as an explosion of product references takes over conversation and determines the concerns and actions of the characters. Moving through such sample stories, or creating such stories herself, the magazine reader learns to follow a narrative thread through or around commercial interruptions, picking her way through an initially unintelligible morass and finding meaning in it. The jolting, discontinuous movement from product to product becomes lively and potentially enjoyable:

As a peace offering he handed her a box of "Dr. Lyon's Tooth Powder" (Dr. Lyon) and a "pocket tin of cocoa butter" (Huyler's) telling her to "look at the brand" (Walter Baker & Co.).

She quickly forgave him and "listened to reason" (Hartshorn). "'Where are the doughnuts?'" (Cottolene) she said, looking around, and then, without giving him time to answer, said, "I want to write for some 'sheet music' (Brehm Bros.) 'With seven months' winter ahead' (American Radiator Co.) I shall have lots of time to practice."[56]

Having gotten to know each other better while shopping, Jim and Miss Cottolene conclude their courtship determined to get positions in an advertisement together. A second example story, "A Leap Year Episode," follows the courtship of Montgomery Ward Smyth and his beloved.

Ladies' World readers responded with some enthusiasm to the challenge of the contest, submitting what the publisher carefully tallied as over eight hundred entries containing a total of 47,821 references to ads.[57] While *Ladies' World* published none of the winning entries itself, an advertising trade journal picked up Beulah Putnam's fourth-prize entry in an article about the contest. Putnam's entry similarly creates a friendly if somewhat bewildering world from advertising slogans and phrases, trailing their required attributions, this time in verse. A courtship and wedding, as in the example stories, provides an occasion for inventories:

They rode on "Rambler" cycles,	(Pope Mfg. Co.)
And heard "Spring Breezes" sigh,	(Anheuser-Busch)
And "Summer Flowers" beguiled them	(Anheuser-Busch)
As idle days flew by; . . .	
Their many wedding presents	
I can't begin to tell:	
"The Wing Piano" furnished	(Wing & Son)
Their tiny parlor well;	
In "this luxurious rocker"	(S. A. Cook & Co.)
They had "refreshing ease;"	(S. A. Cook & Co.)
And "Rogers Bros." gave them	(Meridian Britannia Co.)
A present sure to please.[58]	

Courtship and shopping, marriage and inventories, seem indissoluably linked here. Set the task of constructing a narrative of ad phrases and characters, both the magazine's editors and its readers gravitated toward courtship and marriage stories. The characters, products, and slogans blur in these stories; the characters oscillate between being products of their advertising representatives, like Sunny Jim, and being choosers of products and choosers of spouses, who are themselves products, like Miss Cottolene. While courtship plots offered an accessible set of conventions that might be attractive to a less experienced writer, the ways in which these and other contest stories joined courtship and shopping point to a developing set of fictional conventions that put choosing a mate in the foreground while thematically linking that choice to choosing products. In the contest narratives, however, as if responding to the ad commentators' accounts of the young woman newly minted as a shopper at marriage, courtship and marriage inspire a tour of product names.

Training the Reader

Ladies' World's device of having readers read material closely and then construct a poem or narrative from its fragments was not original with the magazine; the technique had already been used to inculcate familiarity with biblical verses. A six-stanza poem inscribed in a girl's 1835 keepsake book, for example, was constructed from fragments of biblical verse, as the first two verses show.

MARY'S BETTER PART

While many have a name to live	(Rev. 3.1)
Who never felt their plague of hearts;	(1 Kings 8.38)
Give me, O Lord, by faith to know	(Gal. 3.7)
That mine is Mary's better part.	(Luke 10.42)[59]

The biblical exercise taught both its composers and readers that the Bible is a font of language, that the origin of the words of a poem—the fact that they came from the Bible—was at least as important as the content they now embodied. Just as scrapbook compilers adapted skills like cutting and making tasteful arrangements of materials to their work with trade cards, advertisers drew on extant cultural forms and skills to encourage play with advertising. In this contest, *Ladies' World* adapted a form used for ingraining familiarity with the Bible and used it to foster the same level of familiarity with advertising. Readers, who may have known the scriptural exercise, readily applied it to the new material.

Advertising similarly became a font of language. As commodity culture and its advertising provided a nationally intelligible set of references, comparable to that previously most easily found in Protestantism, its promulgators drew on some of the same mechanisms and devices to train people in those references. The ad-identification memory games, for example, echoed popular religious-verse identification parlor games like Bible Salad.[60] Such games and exercises set advertising and religion in the same register: they were both important and worth study. Even those who believed children were inherently drawn to them thought study and direction would reinforce their power. Moreover, both advertising and religion provided a universally available source of reading matter.

When magazines working on behalf of advertisers structured a contest around a familiar Bible-study exercise, they thereby hinted that the advertising itself may have been worthy of reverence. (The notion that the *product* should be honored or even worshiped didn't fully catch on until later, as Roland Marchand demonstrates in his analysis of the product as quasi-religious icon in the advertising of the 1920s.)[61] More centrally, magazines suggested that the gospel of advertising was an appropriate recipient of the same intensive reading practices that had been applied to Protestant scripture study.[62] Because the advertisers and the magazines acting for them were at this point most interested in training readers to read advertising and to think and fantasize within its terms, in nearly all of these contests, the *advertisement*, rather than the *product*, became the focus and object of study.

Within a field in which only a small proportion of companies advertised, virtually any ad was reported to have achieved increased sales in the 1890s. It may have been such reports that fostered advertisers' belief in the overwhelming power of ads and their promises, if the potential customers could only be persuaded to read them. Advertisers alternated between the assumption that readers would find ads intrinsically attractive and the contrary tenet that ads would have to be sweetened in some way to lure readers into taking them in. Scenarios that embedded the product not only in a social world but also within the familiar structures of fiction were central strategies following from this second tenet, as chapter 3 will discuss in more detail. Such advertising sometimes promoted ad reading as much as the sale of the product.

This focus on the ad rather than the product is evident in a pamphlet advertising Coal Oil Johnny's Petroleum Soap, titled *The "Heart Beats" of a Great City*, by Marion Sackett.[63] The pamphlet presents a classic melodramatic tale: a couple has eloped because Elsie's hard-hearted Papa, a widowed "wealthy Boston merchant," has not accepted her choice. Now the couple is in a strange city in a "plain little room . . . as tidy as loving hands could make it"; the husband lies ill of exhaustion and anxiety, and Elsie alone is caring for him. When the doctor arrives, he recognizes from the inscription in a volume of Tennyson in the room that Elsie is the daughter of an old love; he knows her father, too, and believes news of his daughter's situation will soften his heart. But in the meantime, before the doctor begins to search for Papa, Elsie, worried about money, turns from Tennyson to the newspaper and spots an ad: "The manufacturers of Coal Oil Johnny's Petroleum Soap offer a prize of $50 (to be paid within ten days) for the best Poem written on the merits of their Soap, and which can be used as an advertising medium."[64] The company's name and address are given and repeated later.

Elsie, luckily, "had always had a talent for making verses. She *would* try." There's no implication that her ability to write the verse comes from familiarity with the product; rather, she will read the ad. She goes to the store to research it further and is "delighted with the appearance of the soap. It was so purely white and almost transparent, and was perfumed so delicately. How soft it felt to her hands! *Now, she must read what the wrapper said, so she could fully understand what she was to write*, and this is what she read" (emphasis added). The wrapper copy follows. Again, there is no suggestion that she need actually use the product to write reliably about it.

In Elsie's verse that follows, "Coal Oil Johnny's Petroleum Soap, or Blue Monday a Myth," a woman complains of washing clothes. A neighbor offers the soap, saying "Listen while I read what it promises to do." The neighbor says that she, too, was weepy over the wash, but her husband brought her this soap, and washing is now "a *pleasure*." Yet even in the verse, no one is shown actually using the product; rather, they tell one another of it, modeling a new type of conversation and embedding the product within the community. While washing would seem to be the least of Elsie's worries, through involvement with the ad, this lonely, isolated bride invents for herself, and participates in, a new community, in which women look in on one another and pass along information about household work.

Just as Elsie has addressed the envelope, Papa arrives, the husband rises from his sickbed, and the kindly doctor looks on at the reunion. "And now," says the narrator, "I am going to tell you what came for Elsie on the very morning when they were to leave for Boston and *home*."[65] Yes, it was a check for fifty dollars.

Fifty dollars seems a rather high premium for inducing a consumer to read the soap wrapper. But the story embeds mention of the product within a popular form of storytelling and makes the repetition of the product's claims compelling within the story—our heroine's well-being depends on her absorption and use of the claims. The advertising writer, then, writes a story that reenacts his or her own process of ad writing: reading about the product in order to produce an advertisement, rather than using the product and choosing to share the good news of its virtues. Like the *St. Nicholas* contests, the story enlists the reader's sympathies on the side of the ad writer: perhaps all the ads the consumer sees are written by people like Elsie. And the story's structure continually reiterates the value of closely attending to a product's ads and its claims. These substitute for and perhaps even override experience with the product. While the *Ladies' World* contest encouraged readers to encrust a familiar fictional form—the courtship story—with ad and product references, this pamphlet presents a narrative in which ad reading and ad participation become the rescuers, integral to the plot, and demonstrate the heroine's pluck and worth.

The *St. Nicholas* contests, too, suggested that understanding the advertising was more important than having experience with the product. Entrants were to demonstrate that they "*understand the claims that the makers set forth*."[66] To write only "from personal knowledge . . . is to be absurdly scrupulous. . . . The makers of advertisements are but the mouthpieces through which others speak, and their only duty is to help the advertiser in the method of making the suggestions."[67] Like the Coal Oil Johnny pamphlet, *St. Nicholas* contests and the guidance they offered told readers to understand ads as a set of promises. Readers were invited to see ads as fictional structures they could participate in both by creating ads themselves and by accepting the premises of advertising in a register that did not require anyone to compare them with actual goods. Readers were inculcated into the pleasures of advertising as a kind of story-making that animated products and created a mythology separate from any experience of the products themselves. An early, relatively crude version of this story-making took visual form in the St. Nicholas Zoo, which set animal representations of trademarks and other brand references cavorting (see figure 2-5).

Figure 2-5 In this example and other installments, the St. Nicholas Zoo variously animates the trademarks of several brands (the Whitman Chocolate rabbit), puns on brand names, refers to a slogan, or more jarringly refers to the source of products (in another drawing a calf is emblazoned Knox Gelatine and pigs are marked Swifts Hams and Armour's Pork and Beans). Even in this rather crude form, readers are encouraged to imagine products as pleasurably cavorting characters, and to think of ways to include more characters. The pictures, too, primed readers to notice possible references to brand names everywhere, so that the sight of a lion prompted thoughts of Lion Collars or Royal Baking Powder.

Advertising Contests and the Pleasures of Fiction

Readers learned both to play with ads and to look to them for some of the pleasures of fiction. Ads supplied satisfying framing and explanatory structures.
Readers learned to accept that ads were not literal descriptions of the products
(and therefore subject to disappointed expectations) but rather stories about
products in which the reader could participate, either by creating more stories
or by buying the product. Readers were not being encouraged to evaluate the
truthfulness of advertising promises and to resist believing them, but neither
were readers told to believe the ads. Rather, they learned that the issue of
believing or not believing the claims of ads was irrelevant to the pleasure of participating in advertising. Instead, like Elliot Weld Brown, the reader could
embed the product within a complicated interplay of product use, ad-world
play, and fictional construction. Indeed, the product claims Brown advanced—
that Colgate toothpaste is fun to squeeze and tastes like candy—seem the least
important parts of his letter.

One function of magazine ad contests, whether intentional or inadvertent,
was to encourage readers to disregard distinctions between advertising and fiction by inviting readers to fantasize with the materials of the ads and to join ads
to approved worlds of fiction. One competition, for example, invited contestants to use figures from Roman myths to advertise items that had been advertised in *St. Nicholas*; others incorporated figures from children's literature.[68] But
the ads emerged as superior to fiction: they offered options for participation
that the noncommercial stories did not. Eating the cereal that turned Jim
Dumps into Sunny Jim opened a door to the world of this attractive character
that fiction did not offer. It allowed the consumer to be with his or her friend
Sunny Jim, whether or not the cereal fulfilled its implied promise to make the
consumer sunny.

Advertising might offer pleasures not available through fiction, but as *St.
Nicholas's* distinctions between its noncommercial contest (rewarded with
medals, honor, and publication) and its advertising competitions (rewarded
with money) demonstrate, noncommercial creativity was more prestigious than
writing advertising. The publication and honor offered winners of the non-ad
League competitions suggested to winners that they might grow up to join the
ranks of published writers and artists—and, as E. B. White found in 1934
when he went through years of the main St. Nicholas League columns, many
did.[69] White dwells on the thrill of receiving a St. Nicholas League medal, suggesting, for example, that for Edna St. Vincent Millay, a twenty-time St.
Nicholas League competition winner, winning the Pulitzer prize must have
been a letdown after getting her gold or silver League medal as a child.

While advertising contests enticed entrants to harness fiction to advertising, this prestige was threatened when advertising references were added to the
more prestigious, supposedly nonadvertising spheres of fiction. A contest is a
vehicle for this kind of crossover in an episode in Lucy Maude Montgomery's
1915 *Anne of the Island* (a sequel to *Anne of Green Gables*), set around 1900. The

advertising contest gets an aspiring author published and brings her money, but in a setup strikingly parallel to the division of prestige and money in the two *St. Nicholas* competitions, the material reward she receives for her story-turned-ad is repugnant in comparison with the prestige and regard she had hoped to receive for the story in its nonadvertising form. Advertising may have been a space for free play, but it was one that inspired deep ambivalence.

Anne, a college girl, wants to publish a story for the sake of "fame, not filthy lucre, and her literary dreams were as yet untainted by mercenary considerations."[70] The narrator lightly mocks Anne's literary style and approach: Anne's characters have romantic names, and her writing is high flown, full of flowery description, set among city people about whom her friend, Mr. Harrison, complains she knows nothing. In the one homely incident in the story, the heroine bakes a cake. Anne's friend Diana objects that this isn't romantic enough for a heroine, but it's "'one of the best parts of the whole story,' said Anne. And it may be stated that in this she was quite right," the narrator interjects.[71]

Anne puts the story away after it is rejected twice, now agreeing with Mr. Harrison that it's too high-flown. In the chapter "A Dream Turned Upside Down," Anne expresses her revulsion toward a contest she hears of for "the best story that introduced the name of [Rollings Reliable] baking powder," saying, "I think it would be perfectly disgraceful to write a story to advertise a baking powder." Diana soon arrives to tell her that she sent off the story, afraid that Anne had too little faith in it to do it herself, and that it has won: "You know the scene where Averil makes the cake? Well, I just stated that she used the Rollings Reliable in it, and that was why it turned out so well; and then, in the last paragraph, where *Perceval* clasps *Averil* in his arms and says, 'Sweetheart, the beautiful coming years will bring us the fulfillment of our home of dreams,' I added 'in which we will never use any baking powder except Rollings Reliable.'" Anne is appalled, though she has won $25.00—twenty dollars more than the magazine that rejected it would have paid. Anne weeps "tears of shame and outraged sensibility." Feeling disgraced, she turns to her suitor, "What do you think a mother would feel like if she found her child tattooed over with a baking powder advertisement? . . . I loved my poor little story, and I wrote it out of the best that was in me. And it is *sacrilege* to have it degraded to the level of a baking powder advertisement."[72]

But her reaction to the publication and her assumption that advertising is degrading is not shared by her neighbors: "Her humiliation was the consequence of her own ideals only, for Avonlea folks thought it quite splendid that she should have won the prize. Her many friends regarded her with honest admiration; her few foes with scornful envy. . . . Even Mrs. Rachel Lynde was darkly dubious about the propriety of writing fiction, though she was almost reconciled to it by that twenty-five dollar check." For Anne, to have her story published because it succeeds as advertising turns her dream of publication and fame upside down and inverts what her college literature professor has told her: "We were never to write a word for a low or unworthy motive, but always to cling to the highest ideals." Clinging to these high ideals has evidently suggested to the young Anne that an elevated style and subject matter are the appropriate modes of story writing. And yet the "best part" of Anne's story, the homely, realistic, natural moment in which the heroine bakes a cake, is precisely the moment that can be most readily appropriated by advertising. Advertising contests rewarded attention to the everyday universe and to finding or devising new ways to fit commercial references—the new world of brand-name life—into that universe.

The movement of the story into the commercial realm, in which it earns its writer money, paradoxically shifts it into the realm of production in the eyes of her farmer neighbors, who admire and approve of Anne's earning ability. But the insertion of a baking powder reference into her story taints her efforts in the eyes of the standard-setters she tries to impress and bars her from the more prestigious noncommercial realm in which she sought a place when she wrote her story. Although the magazines that rejected her story found her grasp of their standards wanting, the baking powder company presumably recognizes the story as one that joins genteel aspiration with commerce; as revised by Diana, it annexes a conventional romance story to advertising purposes and teaches its readers to expect these genres to cross. It promotes a form of writing in which the arrival of the product may be the story's most satisfying moment, a fictional world which is most familiar and recognizable when it concerns products.

Summary

Advertising's invitation to readers to come play with it and to fantasize using its images and its terms was issued both by the advertisement itself, as we saw in the case of trade card scrapbooks, and later by magazines acting in the interests of advertisers. While an individual advertiser could encourage attention to its own products by issuing attractive advertising cards and even directing collectors to "put this in your album," it could not individually construct the practice of collecting cards and creating idiosyncratic, personal scrapbooks of them. An individual advertiser could run a contest that rewarded attention to the virtues of its own product. But magazines, whose loyalties were to advertisers as a whole as opposed to any one advertiser, could more readily highlight for their readers the pleasures of playing with advertising by directing their play, inventing games that used ads, and rewarding their attention and participation. Advertisers, then, used the institution of the magazine to represent their interests. In this period when editors rather naively assumed that readers saw no conflict between the interests of advertisers and readers, they routinely told their readers how the contests benefited advertisers.

As we saw earlier, the free availability of trade cards, which coexisted with other attractive chromolithographed cards of similar format, encouraged children not only to see these categories as existing in the same register but also to integrate them. Similarly, the advertising contests encouraged readers to bring advertising materials into their lives, to incorporate brand names and advertising slogans into their conversation and writing, and to see the world through a new set of categories. Here, advertising figures became their companions, and advertising could be looked to as a reliable source of cheerful, friendly characters. Its bright and lively sayings were evidently not considered "slang" and therefore not condemned in the middle-class child's household or magazine, in a period in which language of middle-class children was monitored for such lapses. Advertising therefore became an arena of play and pleasure.

In these contests, the magazine overtly offered advertising itself, rather than the thing advertised, as a desirable commodity, something the reader would want to invite into the home and learn more about. Elsewhere, as we will see, magazines used other strategies to make advertising and consumption more central to the magazine reader's experience.

As the conclusion of Beulah Putnam's entry to the *Ladies' World* contest reminds us:

> Now you who read this story
>> Whatever be your name,
> Just study advertisements
>> And "you can do the same."   (*The Saturday Evening Post*)[73]

3

"The Commercial Spirit Ḫas Entered Jn": Speech, Fiction, and Ädvertising

Happy am I as the man who shaves
With Williams' Soap as aid,
And cheery is she who my heart enslaves
As the Pears' "Good Morning" maid.
Her paths shall as easy be, by jove,
As hers who employs Pearline,
And our life shall be one great treasure-trove
Like the back of a magazine.

Edward L. Sabin,
"A Rhapsody in Realism," 1903[1]

The ad. makes the world go round. . . . Take literature. See "Bilton's
New Monthly Magazine." Sixty pages reading; two hundred forty pages
advertising; one million circulation; everybody likes it. Take the
Bible—no ads.; nobody reads it. Take art; what's famous? Gold Dust
Triplets; "Good evening, have you used Pears'?". . . The ad. is the
biggest thing on earth. It sways nations. It wins hearts. It rules destiny.
People cry for ads.

Ellis Parker Butler, *Perkins of Portland*, 1904[2]

The national distribution and advertising of goods by brand name shaped a national culture. In the decades at the turn of the century, proliferating mass-produced goods raised the troubling question whether human relationships, as well as goods, might be reproducible; they inspired anxiety about the extent to which commercial culture in the form of advertising entered into what was classified as properly a purely business-free sphere: the genteel home.

The contests sponsored by magazines and individual advertisers discussed in chapter 2 encouraged readers to use phrases from advertising in their own creations, thereby both fostering the general circulation of brand names and slogans and promoting the readers' more intimate relationship with advertising characters. At the same time—and without overt commercial encouragement—phrases from advertising appeared in stories and speeches, items were mentioned by brand name in novels and entertainments, and characters used in advertising moved into noncommercial speech and writing. National advertising created a cultural shorthand that enabled people across the country to understand a reference to a brand of soap or a joke about an advertising slogan. References to articles in use within the individual and private world of the home—even in intimate use—could be understood nationwide. Then as now, to use an advertising phrase, even among strangers, was to make instant reference to a shared experience.

Such references provided a special type of precision: access to a short-cut realism that allowed speakers and writers to construct pithy, widely understood characterizations. Because one selling point of the new monthly magazines' fiction and columns was their timeliness and responsiveness to current trends and their ability to provide readers with the sense of being in the know, echoes of current advertising slogans were particularly attractive. Such a reference could create an instant air of up-to-date breeziness, as the *St. Nicholas*'s ad manager who persuaded his readers with the pitch, "There are at least a 'Heinz' number of varieties of reasons" in chapter 2 demonstrates. Theater audiences that included recent immigrants as well as longer-established residents could share the pleasure of recognizing a reference to a slogan that they had seen on walls and posters. And yet, of course, ads suggested to readers that they should participate in that national culture as individual consumers.

Slogans and Brand Names in Daily Vocabulary

Besides contributing slogans for a breezier, slangier style, advertising changed the American vocabulary. Samuel Hopkins Adams, famous for his muckraking exposes in 1905 of patent medicines, explicitly praised these changes in a 1909 *Collier's* article on advertising. Adams admired the power of "modern advertising" to enable "the word kodak" to "force . . . its way into our dictionaries"[3] as an exciting new capacity while *Collier's* advertising manager, E. C. Patterson, elaborated this point in the same issue, trumpeting the special triumph of the long-term "publicity advertiser": "By keeping everlastingly at it, he makes his product known and many times a by word: a well known camera manufacturer has advertised his camera so thoroughly and consistently that the word 'Kodak' has actually become synonymous with the word 'camera.' *Advertising costs, but it pays, and the readers reap the benefits.*"[4] According to Patterson, the "benefits" readers could reap from advertising included the availability of new expressions and the pleasure of recognizing them. The knowledgeable *Collier's* reader might recognize Patterson's homage to advertising promoter N. W. Ayer's well-advertised advice to advertisers to "keep everlastingly at it," and thus feel the satisfaction of being among the cognoscenti. In addition, Patterson suggests that a brand-name synonym for a snapshot camera is useful and desirable because it enriches the vocabulary.

Replacing or supplementing generic words with brand names asserted the importance of sorting products into new categories. Advertisers moving into the twentieth century became more interested in cultivating such categorizations; advertising less often promoted unique, novel items and more often positioned a product among competing products by identifying and defining unique qualities in it. So while people had earlier thought in an undifferentiated way of washing, and of the homemade or rough-cut blocks of grocery store soap with which to do it—of soap as a single type of item—now they were encouraged to think of one kind of soap for scouring pots, one for washing cotton clothes and another for wool, and yet another kind for washing hands and body.[5] An 1892 ad for Pears' soap, for example, alerted readers to the dangers of insufficiently differentiated soaps, calling some soaps "quick and sharp . . . the skin becomes rough and tender. . . . Washerwomen suffer severely from soaps no worse than such; indeed the soaps are the same, only one is in cakes and the other in bars."[6]

In her analysis of twentieth-century advertisements, Judith Williamson has noted that promoting and framing a product creates new categories of seeing and new classifications for reading the world.[7] A booklet advertising Hoover vacuum cleaners, for example, names and diagrams five kinds of carpet dirt, defines three as most important and dangerous to carpets, shows how the Hoover removes these three important kinds, and defines those the Hoover can't remove—greasy dirt and stains—as unimportant. Williamson finds that "in every case, the product, whether a cleanser or a kitchen suite, is held out as the "answer" to a problem it claims to solve. . . . The product must distinguish itself from its rivals. And it does this by defining the world around it, creating new categories out of previously undifferentiated areas of experience."[8] Both the newly defined or differentiated problem and its solution, then, were embedded in the proliferating new vocabulary of brand names that replaced the generic or descriptive words for products—so Scourene and Wool Soap, appearing in the late nineteenth century, defined new categories with their names. Trademark law encouraged this process in that it did not extend trademark protection to descriptive words; thus, words used as trademarks could not be merely a descriptive appendage but must be capable of syntactically *replacing* the generic word, as in *Kodak* versus *camera*; *Kleenex* versus *tissue*.[9] A descriptive word could be used if it is arbitrary, as in "Vienna Bread," cited in Herbert Hess's 1915 advertising guide as having been permitted to trademark its name in a court ruling. Even here, though, the arbitrary nature of the name nudged it toward the ability to syntactically replace a generic "I'll have a slice of bread" with "I'll have a slice of Vienna."

In 1915, Hess could consider it an "interesting question" if "Where a coined word has secured such a wide usage as to practically become idiomatic, such as 'kodak,' and 'celluloid,' . . . whether by its evolution into an idiom, the owner is deprived of his legal rights attached thereto."[10] Subsequent court decisions established that a company *could* lose its exclusive rights to a word through precisely the kind of word-of-mouth advertising that would turn its word for a product into not just a household word but the most "natural" and idiomatic word for an object. The conflict is apparent in recent advertisers who both wish to establish their brand names as idioms, or replacements for generic words ("which twin has the Toni?"), and yet must legally protect their exclusive interest in a trademarked word or phrase to avoid having it slip into undifferentiated generic usage, as happened to celluloid, kerosene, linoleum, and aspirin.[11] The more recent legal solution allows companies to trademark entire phrases, such as "Don't leave home without it" or "Please don't squeeze the Charmin." The practical effect of such protection is not to prevent the slogans from becoming popular catchphrases, but rather to announce ownership so as to prevent rivals from using them and to prevent the phrases from entering written language without the tag of the company's ownership. Ideally, from the advertiser's point of view, such phrases roam widely on a secure leash.

In the 1880s and 1890s, however, worry about the loss of a trademark through its use as a generic word was still on the horizon. What was important to advertisers was that it become "natural"—easy and ordinary—to see the world through a new set of categories: not just soap to wash with, but Sapolio or Soapine for scouring, Pearline and Wool Soap for laundry, Pears' or Ivory for hands and body. The skill in categorization praised in nineteenth-century books on collecting, which was cultivated through organizing such materials as trade cards in scrapbooks, was to be brought into play here, too. The enforcement of categories could be deliberate and concerted, as was evident in the *St. Nicholas*'s editor's warning to contestants not to ignore categories: children wouldn't win prizes if they "hint that Sapolio is 'good for soap bubbles' and for 'cleaning silks.'"[12]

Though a magazine's advertising manager might assert with *Collier's* Patterson that readers reaped benefits from this new vocabulary, readers and commentators were not so sanguine about the benefits of advertising's move into daily speech. If advertising language was so alluring and could shape speech so readily, it potentially had power to displace what was seen as the noncommercial speech of human relations and to force speech mechanically into a commercial mold.

Business and Home

Commentary on advertising's move into daily speech suggests that advertising was a problematic point of intersection between the business world and the supposedly noncommercial sphere of the home. In line with the nineteenth-century ideology of separate spheres for men and women, the genteel home was idealized as a refuge from business intrusions, where a properly conducted visitor would not bring up business matters in the parlor.[13] And yet the home, as the site of consumption, was the logical place to direct ad messages. The new ten-cent and genteel magazines introduced ads and commercial references into even the most refined parlors. The Harper's company exploited what it claimed was the special quality of its position in a 1906 address to potential advertisers, touting the ability of *Harper's Monthly*, *Harper's Bazar*, and *Harper's Weekly* to secure social entree for its advertisers to "the class of people who own good homes." Although "the very ones you would wish to reach would not appreciate any direct attempt on your part to interest them in your wares," Harper's asserted, "we should be glad of an opportunity to introduce you to our 'audience,' if you are not already acquainted."[14] Harper's suggested that the suave mediation of its publications and their assurance of entree would sweep the advertiser along without difficulty; it could thus introduce its advertisers into the home on an explicitly social basis. *St. Nicholas* similarly offered to let the advertiser into "the family circle."[15] But advertising materials and references met a mixed reception on their way through the front door.

Ambivalence or antagonism toward this arrival appears even in material from advertisers. The advertising trade journal *Art in Advertising* satirized the possibilities of the intrusion of advertising slogans and language in an 1890 comic scene set in genteel society (see figure 3-1):

> *Miss Fiveinhand (fishing)*: What is so sweet as a day in June?
> *Mr. Pennyscent, who travels for Jones the perfumer (spontaneously)*: You ought to try our new "Tiger Lily" triple extract—wonderful strength and delicacy. A drop placed upon the handkerchief gives the fragrance of June in midwinter.[16]

The satire here is directed at a salesman who too readily falls into his pitch. Elsewhere in the same journal, ordinary customers come in for ridicule for having haplessly absorbed the pervasive Pears' soap slogan "Good morning! Have you used Pears' soap?" A picture of two women talking carries the caption:

TOO FACETIOUS

> *Edith*. It's all over between Jack and me, Penelope.
> *Penelope*. Why, what is the matter?
> *E*. When I said "Good-morning" to him yesterday, what *do* you think he asked me?
> *Pen*. I am sure I don't know, dear.
> *E*. He wanted to know if I had used somebody's soap—wasn't it disgusting? He tried to explain it away afterward, but I told him he had said quite enough.[17]

In England, where the Pears' soap ads originated, one source reports that

> many people did object to the corruption of the familiar "Good morning" greeting. Thanks to a highly successful advertising campaign, it became impossible to use this mundane phrase without fear of the snap response, "Have you used Pears this morning?". . . Still, besides infuriating many people, at least it provided *Punch* with a joke. The scene is the dining room of a country house at breakfast time, the caption, "A Pardonable Lapse of Memory."
> *Hostess*: "Good Morning! Mr Robinson, I hope you slept well."
> *Absent Minded Guest*: "Good morning! Have you used Pears this morning? Er, . . . Oh, yes thank you! Very well."[18]

AT THE COACHING PARADE.

Miss Fiveinhand (fishing): What is so sweet as a day in June ?
Mr. Pennyscent, who travels for Jones the perfumer (spontaneously): You ought to try our new " Tiger Lily" triple
extract—wonderful strength and delicacy. A drop placed upon the handkerchief gives the fragrance of June in midwinter.

Figure 3-1 Advertising pitches intrude into genteel society. (*Art in Advertising* May 1890, p. 39) (Courtesy of the General Research Division, The New York Public Library, Astor, Lenox and Tilden Foundations.)

It was the advertiser's job to penetrate into the social realm and make talk about products, framed within the advertisers' terms, part of ordinary conversation, thus creating word-of-mouth advertising. People sometimes found pleasure in advertising intrusions: as we see in the Pears' examples, repeating advertising slogans was an excuse for making improper comments on other people's bodies or for making overly personal remarks within a frame that legitimated or excused them, as long as other people, unlike Edith, were in on the joke. And yet the jokes also threatened the possibiity that ad slogans might trip people up and cause them to mechanically slip into a commercial message when they intended a social one.

Pears' mimetically illustrated its slogan's ability to project itself beyond the frame in the magazine ad shown in figure 3-2.[19] The picture here makes the printed ad both a barrier to and a doorway into the reader's life, something through which the friendly outstretched hand of Pears' must cross to reach the reader. Just as the hand in this picture iconographically breaks the barrier between the printed paper and the reader, bursting out of the ad and into your life, the advertisement's slogan also succeeded in bursting through the boundary of the ad, crossing into conversation and other spaces. Pears' was one of the

Figure 3-2   This Pears' slogan crosses the border from advertising into conversation, virtually bursting into consumers' lives.
(*Atlantic Monthly*, November 1893, ad p. 38)

earliest companies to advertise in a systematic and organized fashion.[20] Given its
ubiquitous advertising and its slogans' ability to move into ordinary conversation,
it is not surprising that Pears' soap made its way into popular fiction as well.

But brand name references in fiction did more than reflect the larger trend
of brand names' appearance in daily vocabulary. The institution of advertising
told its readers that a unique subjectivity could be expressed through use of
mass-produced goods: it paradoxically instructed consumers to buy mass-produced soap to bring out their individual beauty though thousands of other
women were also buying it. Just as children who made trade card scrapbooks
were learning to express their individual taste by creating individual and
unique arrangements using mass-produced advertising cards, readers of advertising were encouraged to see mass-produced products as something compatible with, and perhaps even the material for, constructing their unique individuality. The tension that people sensed between the mass-produced nature of the
commodity and the claim that it could represent or express their unique subjectivity also appeared in fiction. Concern about the use of mass-produced
commodities is a theme that shadows nineteenth-century fictional references to

brand names. The references participated in the tension between intimate individuality and products available to all; between the idea of the uniqueness of the subject and the ads' proposition that unique subjectivity could be expressed through the purchase and use of mass-produced goods.

Fiction and the Commodity: The Quick or the Dead?

In 1888 *Lippincott's* Magazine brought out Amelie Rives's sensationally sensuous novel *The Quick or the Dead?* In it, Barbara Pomfret, a young widow whose baby and beloved husband Valentine died a year or two before, meets Valentine's cousin John Dering, whose resemblance to her late husband is so striking that she faints when she sees him. She soon comes to love him for his own sake but decides she must not marry him because her own Valentine is waiting for her in heaven and she must remain faithful to him. Much stormy emotion ensues as Barbara repeatedly sends Dering away and calls him back to her Virginia plantation while making her decision.

But in the midst of all this, in a pause between breathless kisses, the lovers have this happy conversation in a strawstack:

> "I love you,—I love you," said each, clinging to the other; and then she settled contentedly down, with her head against his knees, and let one of the greyhound pups curl up between her languid outstretched arms. . . .
>
> "Do you suppose anything ever smelled quite as nice as a strawstack?"
>
> "Yes,—your hair does. What do you put on it?"
>
> "Soap and water."
>
> "Oh, Barbara! do you want me to believe all this is only due to Pears' and your cistern?"
>
> "Indeed, indeed I don't perfume my hair, Jock. I think it's so vulgar. I *hope* it doesn't smell like that!"
>
> "Like what?"
>
> "As though it were all horrid and Lubin's-Extracty."
>
> "That is as original a compound verb as the one Punch's little girl made use of a year ago."
>
> "Oh, yes, I know,—the one about Liebig's-Extract-of-Beefing it. Jock, I think it would be so charming to slide down here together. . . ."[21]

The conversation moves to the length of her hair, and then to more kissing. The scene is a commercial interlude between kisses; it demonstrates that Barbara and Dering read the same magazines, that they inhabit the same world—in fact, a specifically material world, unlike the current realm of the late Valentine Pomfret. The scene focuses on hair, a part of her body which is both within the bounds of acceptable conversation and an emblem of sensuousness (notably so in the period's art nouveau motifs). The use of the brand-named product similarly plays at a boundary: the conversation maintains an intimate and therefore presumably individual focus on Barbara's body and Dering's response to it, while commenting on it in a way that lets the reader participate too, through familiarity with the products discussed.

The novel first appeared in the April 1888 issue of *Lippincott's* following heavy promotion in previous issues, including advertising for the novel and a feature interview with the author, conducted while she, like Barbara, romped with her greyhounds. (*Lippincott's* magazine published entire novels in a single issue, which were then quickly published by the Lippincott publishing house in hardcover.)[22] The novel's large readership and reputation of extraordinary duration rested on its perceived sensuousness. One source reports that it "created an immediate sensation, its young author stepping at one stride from practical obscurity to nationwide distinction. It was more widely commented on than any other novel of the year; it was both attacked and praised, barred from libraries and championed from the pulpit, and within three years, more than 300,000 copies were sold." Reprinted in 1916, it was still on its publisher's active list over ten years later.[23]

Through its many exclamation-marked love scenes ("Jock! kiss me! . . . *Jock!* kiss me! . . . *Jock!* kiss me!"[24]), the only problematic issue between the lovers is whether second attachments are permissible given an implied domestic and materialist version of heaven, where the late Valentine waits for Barbara to join him and resume their marriage. A second marriage for Barbara means bigamy in the beyond. The novel's heaven extends the logic of Elizabeth Stuart Phelps's widely popular 1868 novel of the afterlife, *The Gates Ajar*. But in *The Quick or the Dead?* another element enters: the reproducibility of relationships, the fungibility or interchangeability of people. Even if Dering is not the exact duplicate of Valentine that he first appears to be, he would still replace Valentine as Barbara's husband. Rives's extreme assertion here is that the relationship of people in the roles of husband and wife is so absolutely unique that it can never be reproduced, even in the course of a lifetime. The first love is the authentic one; a second can only be a flawed copy.

The central question of the novel, then, is whether people and relationships are replaceable. Barbara's individual attributes *are* evidently reproducible: any woman who uses Pears' will smell like Barbara. Perhaps everyone or everything *is* reproducible, and Dering can simply find another woman who uses Pears' to replicate his experience with Barbara. The novel resolves this anxiety about authenticity in the face of mass-produced goods in the larger plot, which asserts that people are not interchangeable: even two men who, like Valentine and Dering, are as alike as two cakes of soap, are so authentically individual that one cannot even take on the role that the other held.

In another sense, the Pears' soap passage sidesteps concern with individuality, and specifically with Barbara's reproducibility, by raising the issue of naturalness. The smell of her hair is described as both a product of Barbara's own body and of Pears' soap—that is, of a "natural" world that seems to include Pears'. Dering's strawstack question, "What do you put on it?" picks up on an earlier scene, in which Dering, alone by the fire, becomes conscious of an "elusive perfume": "that exquisitely fresh fragrance which sponges and some women share in common,—a smell of wild grasses and the sea,—of a woman's hair daily washed,—in a word, of Barbara." This smell is present now because earlier she had leaned her head on his breast, and "It was this delicate perfume of her hair which the fire was now drawing from the cloth of his coat." But, ever the gentleman, Dering throws off his coat, because smelling it "was like retaining some spiritual portion of her against her will."[25] Despite the fleeting implication that he could obtain the same rush of memory from sniffing a sea sponge, this scene asserts that this odor is the essence of Barbara, a "spiritual portion of her," authentically and uniquely produced by her and simply unmasked by daily hairwashing.

Despite the fact that Pears' is a scented soap, Barbara is authenticated as a heroine because her smell is seen as not deliberately or artfully applied as perfume would be: like all good romantic heroines, she must be unconscious of her own attractions. Ideally, her smell should be a part of her charm of which she is herself unaware. Certainly she should not have set out deliberately to procure it—although it is acceptable as a by-product of simple hairwashing.[26]

If, on the other hand, the smell of her hair is the product of vulgar perfume or Lubin's Extract deliberately applied, then she would be flawed as a heroine, guilty of having intentionally set out to produce in Dering the attraction about which she claims to be ambivalent. Moreover, if the perfume is all there is to her smell, Dering could replace her with someone else who uses the mass-produced scent. The mass-produced products themselves, however, claimed their own authenticity: ads for the Parfumerie Lubin displayed its trademark and declared that products were "counterfeit if without this stamp," while the slogan for Liebig's Extract of Beef, the third product mentioned in the strawstack scene, instructed consumers to ask for Liebig's and to "insist upon no others being substituted for it."

The novel instructs readers in the products' nuances of signification and teaches the language of distinctions between products, so that the reader learns that smelling of Pears' is better than smelling of Lubin's. By delineating this difference between Pears' and Lubin's the novel adds another layer of social meaning to both products. In using Pears', Barbara, as a discursive creation, takes on the attributes associated with Pears' in the textual realm of its advertising. That is, if we see the novel as in the same register as the ads published in the same magazine, we will learn that as a Pears' user, Barbara should have the "Bright, Clear Complexion and Soft Skin," that Pears' claims to produce; she is presumably one of those with "delicate skin . . . sensitive to the weather," for whom Pears' is specially prepared. She, too, is pure, not "marred by impure alkaline and Colored Toilet Soap." In fact, Pears' associated itself not just with purity and naturalness but with godliness, a matter of concern here given Barbara's interest in joining Valentine in heaven. Its advertising for about a decade featured an endorsement by the Reverend Henry Ward Beecher, asserting "If cleanliness is next to Godliness, soap must be considered as a Means of Grace, and a clergyman who recommends moral things should be willing to recommend Soap"[27] (see figure 3-3).

So the strawstack scene not only features the attractions of the smell of Pears' but also demonstrates the multiple desires that the purchase of a product can embody and allows the reader to participate in both the text of the novel and the text of the ad. Having read Rives's novel, the user can buy the product

Figure 3-3 Henry Ward Beecher promotes cleanliness and godliness. This advertisement appeared frequently in the 1880s, for example in *The Century*, November 1883 and March 1887.

with the belief that it will help her get her man, but at the same time, with the novel in mind, she can purchase an image of herself as the sort of woman who would scorn to buy something to artificially enhance her attractiveness. As promoted in the novel, the soap promises multiple benefits; obliviousness to its benefits happens to be one of them.

This reference to the mass-produced product by name confers a special status on it. We are not simply talking about soap, a generic category into which many things would fit; we are talking about Pears' soap. Barbara is not simply seeking a husband, a generic category into which any eligible man might fit; she is talking about what it means to find another man like Valentine Pomfret and whether it is possible to replace him. Finally, like the dutiful Liebig's Extract of Beef customer, she decides to "insist upon no other being substituted for it" and does not marry Dering. The frightening possibility that the novel raises—that we are not individual subjects with our own souls, but are as reproducible as commodities, and perhaps are ourselves commodities—is thus set aside. Barbara's decision finally affirms that people *are* absolutely individual.

The Quick or the Dead?'s other brand-name scene points to an additional function of the Pears' scene. In dialogue evidently meant to represent conversation among sophisticates, a champagne is mentioned. Dering talks at his New York club with an old Bostonian, Mr. Beanpoddy, who waxes philosophical about love and marriage, opining of weddings that "One decants one's whole life-portion of Perrier-Jouët on that momentous [wedding] morning, and the remainder is apt to become flat."[28]

The Perrier-Jouët reference appears to depend on the reader's recognition that it is champagne (the word *champagne* is not used anywhere in the passage), and might have worked as a snobbish, in-the-know type of reference. Since the reader can work out from context that Perrier-Jouët is a drink that can go flat, however, the reader is given the illusion that he or she *does* possess special knowledge. And since the club world is, in all likelihood, more familiar to *Lippincott's* readers through novels than through experience, the reader is allowed a sense of participation: he or she, picking up contextual clues about the drink, can share in the recognition of upscale names. Moreover, like people playing advertising games who quizzed one another on the slogans and claims of various products, readers of *The Quick or the Dead?* are rewarded for their knowledge of brand names, here with a clearer vision of who the characters are and what sort of world they inhabit.

Similarly, in the novel's strawstack scene, the invocation of a brand name refers past the idyll of the strawstack into the larger, industrial world. It establishes that the characters live in the same world as the reader, despite the novel's exotically heightened emotional terrain. The characters and the reader are familiar with the same magazines and use the same soap, although their knowledge of the products mentioned may well come more from magazine ads than experience.

For the predominantly northern readership of a magazine like *Lippincott's* at this time, the plantation setting was also exotic and signaled romance.[29] A common trope in popular fiction of the 1880s and 1890s positioned the South as a woman, the North as her male suitor; the South as undeveloped and sensuous, the North as industrious and productive, but numb to sensuous pleasures. In numerous stories a hard-driven northern male goes south, often on a holiday, and then, perhaps while recuperating from some injury or illness, falls in love with a southern woman, who embodies a more gracious and sensuous life. This trope appeared often in *Munsey's*, for example, with Frank Munsey's own long-running serial "Derringforth" as just one instance.[30] While John Dering may not be a northerner himself, he lives in New York, and the strawstack scene immediately follows his return from a trip north. (In fact, Barbara remarks mid-strawstack, "Suppose Mr. Beanpoddy could see you now!" marking the distinction between relaxed southern life and the northern world of business duty.)[31] The brand name reference then is perhaps a variant on sensuous southern romping, which here takes the form of enjoying the newly available products.

Of course, the brand name products aren't specifically southern. In fact, the meeting ground for establishing *commonality* between North and South is the world of brand name references, available nationwide, and in the case of the British-made Pears', the French-made Lubin's, and the German-made Liebig's, internationally. Commerce, the national and even international vocabulary of brand names, unites the still markedly separate regions.[32]

Response to The Quick or the Dead?

Reviewers of Rives's novel were sharply divided. Some raved with enthusiasm, hailing Rives for "kick[ing] over the traces of the ordinary vocabulary." "She has a phraseology all her own . . . [of] rich inexactness," wrote *New York World* reviewer Nym Crinkle. Crinkle saw Rives's writing as particularly feminine, operating in a realm beyond masculine understanding, and conflated the writer and her character, speaking of "the feminine sensuousness of the writer and the heroine. . . . You smell the sea taint of her hair and hang on her skirts very much as Dering does." He thus accepted the elision of Pears' and the natural world and even attributed the scent of Pears' in Barbara's hair to the sea. In line with the convention of contrasting the hard-bitten North with the sensuous South, Crinkle found that the book's "chief charm is its absolute freedom from the empirical strait-jacket that Boston has forced with more or less success on the body of contemporaneous American fiction."[33]

Other critics were less enthusiastic. "It is a disagreeable story, fantastically written, and marked by a fleshliness which can only be characterized as coarse," complained one.[34] The heated prose came in for more jovial ridicule in an item in *Munsey's* magazine, which suggested that adolescents found the novel attractively steamy: in a humorously proposed "Misfit Christmas Present Exchange" through which newspaper readers could trade the politely suitable gifts they had received for what they really wanted, an imagined entry announced that a "Young lady of 14 wishes to exchange a wax doll, with real hair, for a copy of

'The Quick or the Dead'; also a rubber cry doll for 25 cents worth of chewing gum, vanilla or strawberry."[35]

While critics didn't directly mention the brand name scenes, they were fodder for mockery in Richard Wrightman's parody of the novel in *The New York World:* "Quick, or You're Dead!: A Modern Novel of Passion of the Most Agonizing Type: The Product of a Brain Which Has Been Warped by Amelie Rives—Sensuous Seances in Old Virginia—the Latest Style of Osculation—A Cyclone of Erotic Sentiment—the Romance of a Liver Pad." In Wrightman's spoof the sexually voracious heroine returns to the mansion where her husband died of exhaustion from her overly ardent demands. Meeting her husband's cousin, she finds he "was like her lost Valorem. He had the Pumpit brow, the Pumpit neck, the Pumpit biceps. Besides, he was a man. She fell to the earth and began to scratch and moan." She attacks him with kisses until he flees, then writes him a note telling him to leave, that his resemblance to Valorem is agonizing to her: "Your midriff flutters exactly as his did. Your electric belt bears the self-same brand."[36]

Wrightman's satire was more sharply directed at Rives's extravagant language and what he saw as Barbara's immoderate sexual appetite than at Rives's use of brand names. But he did link a question important to the novel— whether people are replicable and replaceable—with the idea that identity might be shaped by a mass-produced product. In Wrightman's burlesque, the product is an electric belt, an item promoted as an impotence cure.[37] Here, Barbara's excessive demands evidently produce the same condition in two men, who respond by buying the same device to cure it or to keep up with her; wearing this mass-produced product actually makes Dering more like the lost Valorem. In Wrightman's version, then, Valorem *is* reproducible; neither his nor Dering's qualities are unique. Anyone can generate the same sensation in Barbara by buying the belt. At the same time, another attraction of brand-name talk surfaces here: the existence of these belts and their brands becomes a way to talk about sex, to acknowledge that so many people have the problem (or desire) these belts purport to treat, that the belts, as well as their evidently well-known midriff flutter, are mass-produced and widely available.

Five years after the novel appeared, *Munsey's* pointed to a different significance in Rives's use of a brand name. Teasing that novelists are no longer "improvident . . . victims," *Munsey's* unsigned "Literary Chat" column announced, "The commercial spirit has entered in and made a showing even in these sacred places." The columnist followed up the point with "a recent anecdote from Boston": a novel by a young man, "a graduate of a leading university," was

> accepted by one of the best publishing houses in the literary city. He took the letter of acceptance in his hand and went to a well known champagne firm of advertising proclivities.
>
> "You may see from this letter," he said, "that I am to have a novel published by one of the best houses. It is a society story, and one likely to be read by just the people whom you want to reach. Now what can you offer me to mention your wine as being praised at our best club?"
>
> The champagne dealers probably remembered "The Quick or the Dead" and

its widespread if inadvertent advertisement for a well known soap, and they
offered the young man a pretty sum for a puff of their wine.

But since the bookkeeper for the wine merchant is the brother of the bookkeeper for the publisher, the publisher learns of the scheme, "with the result
that the manuscript was returned to the enterprising author. After this novels
will be edited by that publisher with a view to eliminating advertisements."[38]

The Quick or the Dead? had also of course offered a "widespread if inadvertent advertisement" for Perrier-Jouët champagne, discussed in casual terms at
a club. However, it is the Pears' soap mention, more significant in the novel
because more firmly tied to the heroine's charms, that struck the *Munsey's*
writer as ad-like, as it both drew on and added to the social meaning of Pears'.
The truth or accuracy of this anecdote is less important than the fact that people had noticed how close a mention of a product in fiction was to an actual
advertisement. Moreover, here the business or bookkeeping side of the publisher's work is closely, even literally, related to that of the potential advertiser.
In this anecdote, that very closeness firmly enforces a boundary between editorial and advertising matter, but the story raises questions: What would happen in a city free of the "straitjacket" that Nym Crinkle associates with Boston,
where only the high-priced, non-ad-dependent *Atlantic Monthly* was published,
as opposed to Philadelphia, home of *Lippincott's* and the *Ladies' Home Journal*,
or New York, where most of the new ad-dependent monthlies such as *Munsey's*,
Cosmopolitan, and *McClure's* as well most of the other monthlies that sought
ads, such as *Harper's*, *Scribner's*, and *The Century* were published? What would
happen at a publisher less imbued with the values of literary gentility? One
answer to these questions can be found in discussion of standard magazine
practices in which advertising and reading matter *were* intertwined, chiefly in
what were known as *puffs* or *reading notices*. In these, advertisers, and magazines working for them, deliberately attempted to move trademarks and slogans into everyday speech. The mechanics of puffery show the transmutation
from the sponsored speech of ads to the apparently free—both unfettered and
without cost—speech of conversation.

The Puffing System

Puffing, referred to in the *Munsey's* anecdote, meant touting products or businesses in what appeared to be editorial matter in magazines and newspapers,
and was a long-standing if sometimes controversial practice. Puffs were to be
handled cautiously, according to Nathaniel Fowler's 1897 compendium of
advertising advice. Although it was the publisher's right to offer puffs to advertisers, they should be artfully constructed not to seem like advertisements,
because the reader would feel "illtreated and swindled if more than a reasonable part of that which he has paid for is used indiscriminately for puffing even
reliable dealers." Therefore, "the puff should be newsy, or should give information," with its sales point "disguised as much as possible." Fowler recommended

embedding the mention of the product in a chatty discussion, perhaps about the tastes of people living in different parts of a city for different forms of a product. This "social form of puff writing," would "create a sufficient amount of good-natured talk to assist in selling"; the trick was "to make the puff so interesting and readable that people will read it and not object to it."[39]

As in the *St. Nicholas* contest that encouraged children to make advertising claims the subject of conversational discussion, the advertisers here too wished to infuse conversation with commercial claims in the form of "good-natured talk." The advertiser wished to move discussion of the product as far as possible from its paid-for starting point: the less it seemed like an ad, the greater its value to the advertiser. Similarly, shady financial speculators had been known to parlay cleverly placed reading notices, once they had been quoted elsewhere, into editorial endorsement of their schemes from respected papers, transmuting paid endorsements into freely made recommendations.[40] In Fowler's scheme, however, ads could be transmuted into word-of-mouth advertising, which was most valuable to advertisers: "What, in fact is so good an advertisement as the confidential advice of a woman of influence? . . . Suppose your child is ill, and Mrs. Smith tells your wife that Cantaloupe's food saved her baby's life, a hundred-dollar bill would not be too much to pay for a box or bottle of the food, and so it goes."[41] Reading notices, because they were more like conversation, or like the stories and articles about which people might already talk, seemed to provide a shorter route than straight advertisements through the process of transmuting commercial notices into "good-natured talk."

Although newspapers were more commonly associated with puffs, magazines ran them as well. *Lippincott's*, for example, published a column called "Current Notes" that interspersed nuggets of information about word origins with puffs for patent medicine and pianos, sometimes in the form of anecdotes that only gradually led up to mention of the product. So an 1888 "Current Notes" began by sounding the alarm about the dangers of adulterated foods, leading up to a discourse on the dangers of lime and alum baking powders; it then called it "a suggestive fact that no baking powder except the Royal has been found without either lime or alum."[42] In a period in which the non-puff advertising in all but mail-order magazines was sequestered away from editorial matter and printed separately in the front and back of the magazine, puffs demonstrated the editors' willingness to disregard distinctions between advertising and nonadvertising material. Fiction held a special place, however.

Blind Ads

The *Munsey's* columnist's mildly sardonic reference to heretofore "sacred places," where the commercial spirit was entering, points to the special expectations aroused by fiction. Ads disguised as fiction occupied an uneasy place. Such items, sometimes called "blind ads," might appear in a format similar to other stories in the same magazine. In 1894 the *Ladies' Home Journal* published the full-page story "May's Triumph," in which ailing young Tom is restored to health and falls in love with May, who has taken charge of his cooking. May finally tells Tom's doctor how she did it: "You see, when at home, I studied medicine, and took great interest in matters of hygiene. One day, in reading the 'Medical Times,' I came across the following." A passage by a doctor recommending Cottolene shortening is quoted without indicating whether it is from an ad or an article in the supposed medical journal. May resumes:

> "I made up my mind that Tom was suffering as much from indigestion as anything else, so . . . I suggested preparing all the food for Tom. Knowing that he was passionately fond of pastry, I used Cottolene for shortening, also for frying, and banished lard from the kitchen. Thus it was that Tom's indigestion was cured, and your medicine, which previously had not a fair show, built him up. That's all there was to it."
>
> "No, May," said Tom, "it's not all. Cottolene, as you call it, not only cured my indigestion, making me fit for work, but it gave me the dearest little wife in the world."
>
> "And gave me the best of husbands."[43]

The plot of "May's Triumph" would have been familiar to *Ladies' Home Journal* readers; in a popular formula of the period, a man's temporary disability brings him into a woman's sphere so that he can notice and fall in love with the woman who nurses him back to health. "May's Triumph" was designed to blend into the magazine visually as well; while the alert reader might have noticed that it carried no byline and was set in slightly different type than the rest of the magazine, it was not labeled as an advertisement.

Since advertisements in the *Journal* were already scattered throughout its pages, unlike those in most other middle-class magazines, readers of this magazine were less likely to notice that they were reading an ad before reaching the Cottolene-saturated resolution. But such approaches met with disfavor. A 1902 editorial in *The Independent* declared: "The old fashioned trick advertisement is dead. Some ten years ago the manufacturers of one of the most celebrated breakfast foods advertised by half-column stories, beginning interestingly, but ending abruptly in oatmeal. It was a most exasperating thing, and households and clubs tabooed the brand, in some cases writing to the company the reason why."[44] We have already seen in *Anne of the Island* how easy it was to insert product references, especially at the end of a story. Advertising trade journals satirized such ads, as shown in figure 3-4. One journal published a satirical romance, liberally sprinkled with blanks into which brand names could be inserted. The story ended with a boxing match fought by two suitors in which the underdog triumphed, explaining:

> "You see, dear, for the past week I have eaten nothing else every morning but the celebrated————* breakfast food!"
>
>   * Double rates will be charged for this position, as it is right in the heart of the climax.[45]

Advertising innovator and promoter Earnest Elmo Calkins noted in a 1904 trade journal that "years ago the newspapers used to be full of patent medicine advertising, which started off like a straight news story of the most entertaining character, and wound up with somebody's vegetable compound." But "adver-

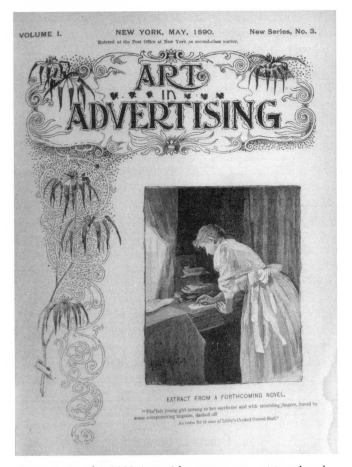

Figure 3-4 This 1890 *Art in Advertising* cover satirizes the ads
that began interestingly but "ended abruptly with oatmeal."
Caption reads: "The fair young girl sprung to her escritoire and
with trembling fingers forced by some overpowering impulse,
dashed off An order for 12 cans of Libby's Cooked Corned Beef."
(Courtesy of the General Research Division, The New York Public
Library, Astor, Lenox and Tilden Foundations.)

tising reached a higher plane, and the business man realized that his actual
business story was the most interesting thing in the world," and so less of such
advertising appeared. In addition, publications had begun to require that such
ads be set in different style from the regular matter, or marked "Adv."[46] Calkins
transmuted complaints about ads that ended abruptly in oatmeal into his complaint that ad references were roughly grafted on to stories. The problem was
not simply that the work had incurred ill will by masquerading as fiction, but
that the elements of the story were separable—Cottolene and Rollings Reliable
were not integral to the narrative.

Yet the possibility of harnessing fiction to advertising tantalized advertisers. In 1900 a writer in an advertising trade journal remarked on a story in an agricultural weekly, in which a turkey was quickly fattened on a diet that included baking powder:

> Of course a fictitious brand was mentioned, and equally of course the publishers would not have admitted the mention of any real brand under any circumstances. Nevertheless, the fact remains that the incidental mention of any brand of baking powder three or four times in the course of the story would have been better advertising than almost any quantity of the so-called blind ads . . . and quite as good as several times the same space in the regulation form.[47]

Perhaps, the writer suggested, if the story were absorbing and the brand were mentioned throughout, readers would actually be interested in the way it is advertised, and therefore interested in the product. He suggested inserting such stories in the advertising pages immediately following reading matter—in essence continuing to disguise the ads as stories, but integrating the product throughout the story instead of ending abruptly in oatmeal.[48]

Beyond this tentative straddling of the physical boundary between ads and reading matter enforced by the magazines, more incidental—or seemingly incidental—literary mentions of salable items succeeded as advertising. The novelist Hall Caine's vivid description of the Isle of Man had raised Manx real estate prices, according to an 1899 advertising trade journal report. The journal saw his fiction as an instance of "cloth[ing] these splendid advertisements with interesting romances," which the real-estate buyers presumably assumed would come with the property, thus raising property values.[49] Similarly, Calkins reported in 1904 that *The Lightning Conductor* told of a trip through southern France in a Napier automobile: "The story itself is worth reading, both for its exciting plot and for its vivid picture of the country through which the tour was made. In addition to this, it sets forth in a most delightful way the real pleasures of an automobile which of course settle around the well-behaved Napier."[50] Unlike Pears', which seems not to have exploited the mention of its product in Rives's novel, the Napier company placed ads that specifically referred to the book.[51] Consequently, the book bore extra-literary fruit, first stimulating sales of the car in England, and now, "this book was published in this country, was widely read, and made a big hit, and as a natural, logical result of that circulation an agency has been opened in this country for the sale of the Napier Car."[52] While in this pioneer motoring period the Williamsons would not have been likely to omit mention of the make of their auto, such mentions continued to jar in other contexts.

Brand Rames as Satiric Targets

Brand-name references were useful to fiction writers other than Amelie Rives, but often for different reasons. Satirists found brand names to be convenient targets, as in Richard Wrightman's mockery of brand-name references in his parody of *The Quick or the Dead?*. Other writers used brand-name references to criticize an activity as overly commercial or to tag a character who glibly speaks in brand names and slogans as too concerned with commercial matters. In William Dean Howells's 1890 novel *A Hazard of New Fortunes*, Fulkerson, the business manager of the fictitious magazine *Every Other Week*, is characterized as "a pure advertising essence." Fulkerson celebrates the popularity of his new magazine in brand-name references and in terms derived from advertising slogans: "It's the talk of clubs and the dinner tables; children cry for it; it's the Castoria of literature, and the Pearline of art, the Won't-be-happy-till-he-gets-it of every enlightened man, woman, and child in this vast city."[53]

Fulkerson's adaptations of slogans (such as Castoria's "children cry for it," and Pears' "won't be happy till he gets it") along with his use of the brand names Castoria and Pearline, characterize him as comically steeped in advertising. Similarly, in Ellis Parker Butler's Perkins of Portland stories (satires written in the 1890s and 1900s), avid huckster and advertising promoter Perkins's speech occasionally veers between real and parodied advertising slogans. Explaining to his narrator sidekick that he is from neither Portland, Maine, nor Portland, Oregon, he proclaims, "It's all in the ad. 'Mr. Perkins of Portland' is a phrase to draw dollars. I'm from Chicago. Get a phrase built like a watch, press the button, and the babies cry for it," thereby combining the slogans of Sterling bicycles, Kodak cameras, and Castoria castor oil into one insistent self-advertisement.[54]

In fiction, it was the advertising practitioners such as Fulkerson and Perkins who seemed most susceptible to embellishing their speech with advertising slogans, as did real practitioners such as *St. Nicholas*'s advertising manager with his "Heinz number of varieties." Howells's publisher, in a full-page ad for *A Hazard of New Fortunes* in an advertising trade journal, assumed that Fulkerson's speech would resonate with real advertising workers: the ad quoted the "children cry for it" passage, along with others, and called Fulkerson "probably one of the most interesting characters of fiction produced by the brilliant group of writers of which Mr. Howells is the acknowledged leader."[55] In contrast, Ellis Parker Butler represented advertising slogans as attractive to a wider group of speakers than ad practitioners. In one story, Perkins advertises with the slogan, "Perkins pays the freight," persuading customers to buy large bottles of mineral water. As the product catches on,

> We . . . took up magazine advertising on a big scale. Wherever man met man, the catchwords, "Perkins pays the freight," were bandied to and fro. "How can you afford a new hat?" "Oh, 'Perkins pays the freight'!"
>
> The comic papers made jokes about it, the daily papers made cartoons about it, no vaudeville sketch was complete without a reference to Perkins paying the freight, and the comic opera hit of the year was the one in which six jolly girls clinked champagne glasses while singing the song ending:
>
> > "To us no pleasure lost is,
> > And we go a merry gait;
> > We don't care what the cost is,
> > For Perkins pays the freight."[56]

Butler's lively slogan is both heavily advertised and somehow intrinsically irre-
sistible, fulfilling a need, or at least an opportunity, for a slogan to catch the
popular imagination.

Butler showed vaudeville magnetically drawing advertising catchphrases
into its lively miscellany, while in a later episode Perkins declares the more staid
legitimate theater moribund because it *lacks* "the life-blood of today": advertising.[57] In a claim more grandiose than that of the most zealous advertising promoters of his time, Perkins extrapolates from the economics of magazine production the claim that advertising is the central driving force of the entire
economy: "Why . . . do men make magazines? To sell ad. space in them! Why
build barns and fences? To sell ad. space! Why run street-cars? To sell ad. space!
But the drama is neglected. In ten years there will be no more drama." With this
in mind, he commissions an entirely formulaic musical, *The Princess of Pilliwink*,
purely to sell ad space in it: "What's a magazine? So many pages of ad. space.
What's a play? So many minutes of ad. space. Price, one hundred dollars a
minute. Special situations in the plot extra." By the time Perkins is finished
working on it, the play has "some modern interest. . . . It went right to the spot."
Unlike the stories that began interestingly only to disappoint their readers by
ending abruptly in oatmeal, Perkins's play charms his audience by its modern
involvement with advertising throughout:

> There was a Winton Auto on the stage when the curtain rose, and from then until
> the happy couple boarded the Green Line Flyer in the last scene the interest was
> intense. There was a shipwreck, where all hands were saved by floating ashore on
> Ivory Soap,—it floats,—and you should have heard the applause when the hero
> laughed in the villain's face and said, "Kill me, then. I have no fear. I am insured
> in the Prudential Insurance Company. It has the strength of Port Arthur."[58]

The play provides topical interest by changing its ads on a monthly basis, with
scenes rewritten to fit. When food producers predominate among the advertisers, and the cast threatens to quit to avoid eating "thirteen consecutive breakfast foods," Perkins makes his play even more popular by cutting the eating
scenes and inviting the audience up to sample the foods after the play. The play
now begins with autos and ends logically, rather than abruptly, with oatmeal,
thereby delighting its audience. This move suggests one attraction that con-
temporary ad references embodied and that Butler's advertisement-avid crowds
sought: ads and ad references themselves offered vicarious consumption. The
fictional audience watching the actors ride the Winton Auto not only got an
outline of the car's social meaning as they saw it embedded in a larger script
(even one as formulaic and improbable as *The Princess of Pilliwink*), but also
were allowed to vicariously participate in the use of the product.[59]

The advertising scheme that Butler sketched in *The Princess* anticipates
later radio and television commercials in which products are in fact displayed
through dramatic scripts. It still more markedly prefigures late-twentieth-century product placements in films and television, a practice in which advertisers pay to have their products shown in use in a production.[60] In addition to
the added economic incentives, product placement has obvious attractions for

a writer working in a literal visual medium like film, television, or naturalistic theater. If characters eat cereal in a household where the box would characteristically be on the table, there are essentially three choices: either the scene has to be carefully arranged to avoid showing the box; the cereal must be an invented, nonexistent brand; or the box must be an identifiable brand of cereal. Careful exclusion of the box from view is cumbersome, and an invented brand calls attention to itself as a creation of the author and interferes with the premise of a "realistic" scene, the premise that the reader and the characters exist in the same universe. On the other hand, as with Rives's use of Pears' soap, the appearance of the brand-name item reinforces the reader or audience's sense of being in the same universe as the characters, of shopping for the same products, reading the same magazines. The choice of brand itself becomes a source of information about the characters, about their tastes, and about their class position, just as the Marches in *Hazard of New Fortunes* learn about the other characters through their choice of home furnishings.

In literature as well, brand names became a convenient, nationally intelligible shorthand for conveying information about characters and their tastes. Amy Kaplan has suggested that realist novelists such as Howells "at once resist and participate in the domination of a mass market as the arbiter of America's nation idiom,"[61] but it's worth noting not only that Howells's contemporaries saw in Fulkerson realist accuracy rather than satire and resistance, but that the use of brand names itself was labeled realism. Thus, the satire "A Rhapsody in Realism" in an ad trade journal:

> Oh, my love has regular Sozodont teeth,
> And a Packer's cloud of hair;
> It ripples down to the floor beneath
> Like that of the damsel Ayer.[62]

But for genteel nineteenth-century readers, brand-name realism was too realistic, an orientation that may explain the otherwise improbable complaint of one reader of the *Quick or the Dead?* that the novel embodied "realism with a vengeance."[63] It provided too accessible a shorthand, a shorthand legible both to genteel and nongenteel readers. Just as in a genteel household the cereal box would not be brought to the dining table, elite space and culture were marked by the deliberately obtained freedom from advertising references. Brand-name references thus came to mark a work as low-culture. The elite therefore came to object to nonsatiric advertising references because they suggested participation in the nonelite, low-culture world.

Harper's, *St. Nicholas*, and *McClure's* were among the magazines that offered to serve advertisers by bridging this gap by volunteering to introduce advertisers to their readers. The fiction and other material these magazines carried would serve to legitimize and gain entree for the ads, which could then be allowed into the parlor.

Appropriating High Culture in Advertising

The advertiser was an interloper, a member of the rising class. Advertisers'
attempts to annex high-culture cachet to their products and to have their products cross the boundary between advertisement and other forms of cultural
display produced another variant of the anxiety generated by advertising's
intrusion into provinces presumed to be free of commercial taint, but which are
themselves commodities.

Advertisers were attracted to the possibility of using the work of prominent
artists in ads, in part because the artist's name itself would serve as an endorsement. An advertising trade journal, for example, in 1903 stated that it "is taken
for granted that no artist of reputation will make an illustration for use in connection with any commodity or enterprise that is not above suspicion or taint
in every particular."[64] The writer's assumption here is that the artist serves as a
priestly definer of a sphere of beauty and truth. Naturally, even before the fad
for celebrity endorsements took off, someone who could put such an imprimatur on a product seemed a desirable adjunct to the advertising project.

Pears' soap was particularly enterprising in annexing high culture to its
advertising. As an early and aggressive magazine advertiser, it regularly ran full-
page ads in most of the monthlies. And as we have seen, it associated its product with priestly pronouncements on cleanliness and godliness, and appropriated beauty and truth via endorsements from opera singers like Adelina Patti.
Pears' signal move into the arena of high culture was its purchase of Royal Aca-
demician John Everett Millais's 1885 painting "A Child's World." The painting,
which showed a little boy looking up wonderingly at soap bubbles, became
known in Pears' reproduction as "Bubbles" (see figure 3-5). It was first bought
and reproduced by *The Illustrated London News*, which then sold it, along with
its copyright, to Pears'. The company initially produced high-quality reproductions devoid of any advertising associating it with Pears' soap. In subsequent reproductions, however, the name of the company was displayed above
the image, and a bar of Pears' soap was added at the child's feet; when it
appeared in magazine ads other copy was added.

Although Millais was evidently not party to the sale to Pears', he came
under attack for what was seen as a "degradation," or selling out, of his art.[65] It
disturbed his fans that a picture they cherished now carried commercial associations. The novelist Marie Corelli, for example, attacked Millais in one of her
novels for supposedly "painting the little green boy blowing bubbles of Pears'
soap." She was rebuked by the artist, and replied with a letter articulating her
pained response to the commercial use of the painting:

> I get inwardly wrathful whenever I think of your "Bubbles" in the hands of Pears
> as a soap advertisement. . . . I have seen and *loved* the *original picture* . . . and I
> look upon all Pears' posters as gross libels, both of your work and you. . . . Now
> the "thousands of poor people" you allude to are no doubt very well-meaning in
> their way, but they cannot be said to understand painting; and numbers of them
> think you did the picture solely for Pears, and that it is exactly like the exagger-

Figure 3-5 Pears' "Bubbles," originally Millais's "A Child's World," which ran as a full-page ad in numerous magazines. Pears' added the bar of soap at right. The caption boasts of the company's ownership of elite artwork. (*Atlantic Monthly*, March 1888, p. 32.)

ated poster. . . . "Bubbles" should hang beside Sir Joshua's "Age of Innocence" in the National Gallery, where the poor people could go and see it with the veneration that befits all great Art.[66]

Evidently responding to Millais's defense that Pears' reproduction allowed "thousands of poor people" to have copies of the painting, Corelli seems to veer between distress that poor reproduction quality exaggerates and distorts the painting and concern that other people will make the same assumption about Millais's motive in creating the work that she did: that he painted it as an ad. She looks to institutions to reassert boundaries between high and low art: she finds it acceptable for poor people to enjoy the painting, as long as they do so within an institution of high culture like the National Gallery, where their veneration would be enforced. Corelli argues that if people own their own exaggerated reproductions of the painting, they could fail to properly venerate it. Like the scrapbook compilers with their chromos, such owners could more freely manipulate it themselves and theoretically find their own (incorrect) meanings in it.

Corelli was herself a notoriously advertising-dependent author (*Munsey's* refers to her as "the gifted press agent and novelist").[67] Ironically, while she expressed revulsion at the commercial use of the Millais picture, she also referred to it by the name Pears' gave it. Moreover, long after the picture had become associated with the soap, she mentions in the same letter that she'd had her child pose in a public tableau vivant of "Bubbles." This apparent contradiction implies that visual references to the painting *could* still stand on their own, with their reference to the more familiar ad muted. Nevertheless, the use of "Bubbles" in a tableau capitalized on the publicity that had made the picture familiar enough to be instantly recognizable—like the Detroit tableaus of Spotless Town and the Pears' soap lady.

Summary

Pears' use of "Bubbles" again held forth the tantalizing possibility that respected art work and respected artists could serve advertising. But it was one thing to sketch a bar of brand-named soap on the bottom of a famous painting of a child blowing bubbles, thereby converting the entire scene into a visual narrative of product use so that he becomes "the little green boy blowing bubbles of Pears' soap." It was another matter to more thoroughly blend Cottolene and brand-named baking powder into a convincing verbal narrative mix.

Trade cards and posters took advantage of the instantaneous visual impact of such images, but magazines were less successful. Although the technology for reproducing illustrations was improving rapidly, advertising texts and trade journals often warned advertisers against running pictures in magazines and newspapers because of the difficulty of controlling the quality of reproduction. In the verbal realm, blind ads tried to hide the mention of the product in a story, and the advertisers paid the magazine to publish such an ad in an ambiguous place, on the borderline between advertising and editorial matter. Advertisers considered puffing, or the insertion of what both advertiser and publisher recognized as disguised ads, desirable, though full of pitfalls. Another way of integrating ad references into verbal narrative was the substitution of brand names for generic words, so that characters spoke of Pears' and not of soap, of Rollings Reliable and not of baking powder. This had attractions as a type of lively and up-to-date speech, but as we have seen, writers who employed this method tended to draw unfavorable notice. While the most successful instances of such brand-name use were not deliberately constructed as ads, advertising ultimately learned to construct ads that echoed their ability to feature the product's attractions while demonstrating the multiple desires that the product could embody.

It was this kind of promotion of a product name—as in *The Quick or the Dead?*, or of a locality, as in the Manx real estate account—that anticipated later advertising strategies of product mentions embedded in social scenarios. These scenarios didn't simply show the product itself, or tell that it is useful or "the best," as earlier magazine ads tended to do, but rather demonstrated the uses of the product and its impact among people. As one commentator remarked in 1902, readers must not only see but also remember a product like a suit or a shoe, and this could best be brought about by "putting the suit in a drawing room scene or the shoe on the foot of a handsomely-dressed woman."[68] Moving the shoe or the suit into relationship with people, discussing or illustrating its social meaning, was the strategy increasingly pursued by advertisers. In other words, advertising learned from fiction.

Magazines enlarged the frame in the project of constructing ad scenarios that could shape how people understood the social uses of a product. They provided an expanded venue in which reference to commodities was given new context and a new social meaning. Their loyalty to many advertisers meant that such references were rarely to a single brand-named product. Instead, the magazine could frame a new context for a commodity. It did this perhaps most successfully when the commodity itself was a new one, and its appearance in the magazine therefore bolstered the magazine's claims to timeliness and currency.

4

Reframing the Bicycle: Magazines and Scorching Women

Riding the wheel, our own powers are revealed to us, a new sense is seemingly created. . . . You have conquered a new world, and exultingly you take possession of it.
                                          Maria E. Ward, *Bicycling for Ladies:*
                                          The Common Sense of Bicycling (1896)[1]

It would certainly not be desirable for a young woman to get her first ideas of her sex from a bicycle ride.
                        H. O. Carrington in *The American Midwife* (1896)[2]

When the safety bicycle in the 1890s made bicycling accessible to women, wheelwomen found themselves riding through contested terrain. The bicycle offered new mobility: new freedoms that attracted feminists and other women, yet made it the target of conservative attack. Both defense and attack took medicalized form: anti-bicyclers claimed that riding would ruin women's sexual health by promoting masturbation, and would compromise gender definition as well. Pro-bicyclers asserted that bicycling would strengthen women's bodies—and thereby make them more fit for motherhood. Such claims are familiar from a period in which many discourses were cast in medical terms and issues as diverse as shoplifting and women's education were tied to reproductive health.[3] These conflicts unfolded in the world of commerce, as commercial interests negotiated with and within shifting ideas about women's bicycle riding. Specifically, the discourse of consumption constituted by the advertising, articles, and fiction within the developing mass-market magazine of the 1890s subsumed both feminist and conservative views in the interest of sales. In effect, these advertising-dependent magazines asserted a version of women's bicycling that reframed its apparent social risks as benefits. The marketing of bicycles and its relationship to the magazines offers a striking example of the interplay of fiction and advertising in the new magazines. The editors' and publishers' sense that they had interests and goals in common with "progressive" advertisers helped shape this marketing, in all likelihood without overt pressure from advertisers.

Although magazine ads themselves in this period sought to persuade readers chiefly by attracting attention to their products and by verbally offering rational and quasi-rational appeals, complicated narratives that embedded products in a social context and associated them with romance, happiness, freedom, social acceptance, and socially approved behavior did appear in the 1890s—in fiction that appeared to serve no commercial purpose. Such fiction prefigured later, more deliberate advertising strategies while it accustomed readers to appreciate and respond to the kinds of narratives that advertising later used more extensively. Unlike the blind ads and the stories that ended abruptly with oatmeal, the successful product-oriented stories did not endorse individual brands, but promoted product categories. The mechanism of the advertising-dependent magazine harnessed this ability in the interests of advertisers. In the case of bicycles, magazine fiction allayed a specific set of concerns about the mobility of women on bicycles and reframed women's bicycling as a trip that would end in married happiness.

Gender and Bicycles

The safety bicycle—that is, the bicycle roughly as we know it now, with equal-sized wheels and inflated tires—was a popular novelty in the 1890s. It opened up to the middle class kinds of travel that had previously been available only to those wealthy enough to keep a horse, while it posed new problems and opportunities for its makers and marketers. Its predecessor, the "ordinary" or high-wheel bicycle, with an enormous wheel in front driven directly by the pedals at its axle and a single small wheel in back, was dangerous and difficult to ride. Notoriously, a bump in the road could send the rider over the handlebars. It was impossible to ride in skirts, even divided skirts, and riding it was specifically marked as a masculine pursuit.[4] Riders were a relatively small group of athletic young men.[5]

The high-wheeler was put to use in already-established fictional formulas that placed the man in the world and the woman at home. One such general formula appeared frequently in *Godey's Lady's Book* stories in the 1880s; it parallels the formula of the blind ad "May's Triumph" discussed in chapter 3. In it, a man is knocked out, or temporarily disabled, so that he may become a suitable husband. His injury obliges him to stay in one place while he sees how well some woman takes care of him ("You have been ill—very ill. . . . The doctor says you are not to lift your head for several days"[6]) and has a chance to fall in love with her. This formula enables courtship in a world in which women lived circumscribed lives, their accomplishments largely invisible, while men enjoyed freedom of movement. Retrieved from the highways and deposited in a woman's household, the man is kept still long enough to notice the tender womanly and housewifely arts she has cultivated.[7]

The high-wheel bicycle is appended to this formula as a handy emblem of male mobility in the 1888 *Godey's Lady's Book* story, Max Vander Weyde's "A Turn of the Wheel." In it, a high-wheeler both represents and enables a man's mobility; it opens up worlds into which he would not otherwise be admitted. Here, a young American wheeling through Provence is distracted by the sight of a young woman standing in a doorway, who has "the unconscious trick of posing like a bisque statuette." He takes a header over his handlebars and must be brought in and nursed, although Americans are normally forbidden in this household (the statuette's father disappeared fighting on the Union side in the Civil War). In this story even the high-wheel bicycle's drawback, its tendency to throw riders over the handlebars, becomes a useful feature: the rider falls off his bicycle into a household where he then falls in love. Later, evicted by an angry uncle, the American rides off to Paris and locates his love's long-lost father, who gives the couple permission to marry. The man must be knocked off his high-wheeler to bring him within a woman's sphere and make him marriageable: otherwise he would go by too fast.

As long as such a man was on a high-wheeler, he was definitively outside the sphere of women. The safety bicycle's accessibility to both men and women, however, raised new social issues. Since riding a safety offered an enhanced version of the freedoms that riding a high-wheeler had allowed men, the act of riding, as well as the safety bicycle itself, was seen as essentially masculine. Women's riding therefore posed a threat to gender definition. It was also perceived as threatening women's sexual purity, as will be discussed later in this chapter. And when unmarried men and women rode together, cycling threatened chastity and order.

Manufacturers wished to sell bicycles to as many people as possible—both men and women. The safety bicycle, though apparently nongendered, was understood to be masculine, so women's riding had to be made socially acceptable to sell safety bicycles to a larger market. Assurance that riding the new safety bicycle was appropriate to their gender was therefore important to potential female buyers and helped allay both their own concerns and those of others who might object to their riding.

Manufacturers pursued a variety of strategies for gendering the safety bicycle, asserting that a women's mode of riding was available and that riding need not be masculinizing. They quickly differentiated models of bicycles for men and women: the diamond-shaped frame (similar in shape to present-day men's bicycles) was presented as standard; the drop frame that allowed riding in a skirt became the marked category, the women's version. (Because the diamond frame was structurally stronger, reinforcing hardware often added as much as ten pounds to the weight of women's models.) The notion that bicycles should be gendered soon extended beyond accommodating dresses. An 1895 advertisement for Columbia bicycles showed special diamond-frame women's bicycles for women who planned to ride in "zouave [bloomer] or knickerbocker costume." Although the bicycle in the ad appears no different in size or shape from the men's models, the manufacturer gave it a distinctive name and sold it as a woman's bicycle. Through its design, the ad itself, in the form of a paper doll, asserted the gender-appropriateness of women's riding and simultaneously reformulated questions of appropriate dress as questions of style and fashion (see figure 4-1).[8] Manufacturers assigned names to the different models to emphasize their genderedness: the ambassadorial Envoy for men, the birdlike Fleetwing for women, the Napoleon and Josephine, and the Victor and Victoria, for example.[9]

Advertising the Bicycle

Other forms of advertising linked bicycles with codes of femininity. Bicycle manufacturers were, according to one advertising historian, "trail-blazing pioneer[s]" of magazine ads in their extensive use of space.[10] Designed by artists such as Maxfield Parrish and Will Bradley, the visual elements of the graphically sophisticated bicycle ads invested bicycles with glamour and the graphic associations of up-to-dateness. Art nouveau's flowing, curvilinear organic forms linked bicycles with the natural world and echoed a theme of numerous magazine articles of the period in which bicycles provided city dwellers with easy touristic access to a countryside appreciated as scenic and picturesque. Although articles on bicycling into the countryside typically discussed the technicalities of the tourists' arrangements and issues of bicycle repair, traveling by bicycle was nonetheless framed as a "natural" experience, freeing the urbanite from such "mechanical" constraints as railroad schedules.[11] Bicycles themselves became design elements within the advertising posters and magazine ads, and the viewer was invited to become part of an edenic, harmonious, or exotic picture by buying and riding the bicycle (see figures 4-2 and 4-3). In these advertisements, as in the articles, the bicycle is imagined not as a machine in the garden but as part of the garden.

While cartoons in humor magazines of the period such as *Life* and *Punch* satirized women bicyclists as mannishly dressed menaces, and a few bicycle and accessory manufacturers' ads displayed very young women riding in daringly short skirts, the more naturalistic of the drawings in ads and catalogs provided constant visual reassurance that women could ride the bicycle with grace and even modesty.[12] Such advertising art frequently showed women riding in skirts that covered their feet as they rode, and which would appear to drag on the ground if the figure in the drawing stood (see figure 4-4). Both the naturalistic drawings and the stylized art nouveau productions strove to demonstrate that bicycling women could be both decorative and decorous. The Victor ad in figure 4-2, for example, highlights the curves of the drop-frame on the middle bicycle and uses the netting of the rear-wheel dress guard on the furthest bicycle as an ornamental embellishment while emphasizing the three women riders themselves as flowing decorative elements.

Figure 4-1 Paper doll for the Pope Manufacturing Company. Pope drew on and encouraged interest in new forms of dress for bicycling with a series of paper dolls wearing bicycling costumes designed by well-known couturiers. Each doll stands with a Columbia bicycle. (Courtesy of The Winterthur Library: Joseph Downs Collection of Manuscripts and Printed Ephemera.)

Figure 4-2 This 1896 ad for Victor bicycles drawn by Will Bradley appeared in many magazines, including *The Century* and *Harpers*. (*The Century*, February 1896, back cover)

The ads thus visually proclaimed a suitable women's mode for riding, on a suitably differentiated bicycle. The written copy of 1890s bicycle ads was less sophisticated than their graphics, however. Aside from fanciful and suggestive model names, such copy often restricted itself to informational claims about price and speed. But unlike the narrative blind ads and more in line with other magazine ad copy of the period, bicycle advertising copy rarely suggested social consequences following use of the product. That was to be left to magazine fiction.

Figure 4-3   This 1896 ad for Victoria bicycles entwined even a bloomer-clad rider in the harmonious garden. (*Ladies' Home Journal*, March 1896, p. 29)

The Threat of the Mobile Woman

For women in particular, the new mobility the bicycle allowed offered freer movement in new spheres, outside the family and home—heady new freedoms, which feminists celebrated.[13] Suffragist and temperance leader Frances Willard called her bicycle an "implement of power" and "rejoiced in perceiving the impetus that this uncompromising but fascinating and illimitably capable

Figure 4-4 Drawings in advertising material often showed women riding in improbably long skirts. ("My Morning Spin": *Ladies' Home Journal*, November 1892, p. 20; Ladies' Rambler: *Ladies' Home Journal*, May 1892, p. 13) (Both courtesy of the General Research Division, The New York Public Library, Astor, Lenox and Tilden Foundations.)

machine would give to that blessed 'woman question.'"[14] Maria E. Ward's 1896 manual *Bicycling for Ladies* hints at a connection between cycling and suffrage, linking riding to both autonomy and responsible citizenship:

> Riding the wheel, our own powers are revealed to us. . . . You have conquered a new world, and exultingly you take possession of it. . . . You feel at once the keenest sense of responsibility. . . . [Y]ou become alert, active, quick-sighted, and keenly alive as well to the rights of others as to what is due yourself. . . . To the many who wish to be actively at work in the world, the opportunity has come.[15]

The "New Woman on a bicycle . . . exercised power . . . changing the conventions of courtship and chaperonage, of marriage and travel," Patricia Marks notes in her survey of the period.[16] To some, the bicycle itself seemed to offer a transcendent solution to women's problems. As playwright Marguerite Merington wrote in a 1895 *Scribner's* article, "Woman and the Bicycle," "Now and again a complaint arises of the narrowness of woman's sphere. For such disorder of the soul the sufferer can do no better than to flatten her sphere to a circle, mount it, and take to the road."[17] Similarly, for Frances Willard, the bicycle was a means of access to a larger world, and even became that world itself: "I began to feel that myself plus the bicycle equaled myself plus the world, upon whose spinning wheel we must all learn to ride." She explained in her account of learning to ride that she took up bicycling at age fifty-three in part so her example would "help women to a wider world."[18]

Conservatives attacked women's bicycling in correspondingly hyperbolic terms, as a force that would disrupt social roles by allowing women freedom of movement beyond family surveillance and outside traditional gender roles. Like other attacks on women's opportunities for autonomy in this period, this one was framed in sexual terms. One way conservative fear of the disruptive potential of women's riding was articulated was through medical discourse that attacked bicycling as both masculinizing and a threat to sexual purity.

Pro-bicycling forces also used medical discourse to endorse women's riding. They shared with the anti-bicycling opposition the notion that medical pronouncement on women's bicycling was necessary and called for, that pleasurable physical activity undertaken by women should come under medical authority. Nearly every book on bicycling from the period included a discussion, citing medical authorities, of the effect of bicycling on women's health. Special magazine issues on bicycling routinely included articles from physicians commenting on riding and women's health.[19] Even Frances Willard followed this convention, interrupting her discussion of her own riding experiences to address what she believed to be an issue worrying the reader: "And now comes the question 'What do the doctors say?' Here follow several testimonials."[20] The testimonials she chose unsurprisingly favored women's riding; their inclusion at all, however, demonstrates the pervasive assumption that medical sanction for or against women's riding was appropriate and even necessary.

As we will see, both sides of the specifically medical anti-bicycling discussion had in common an interest in keeping women within traditional roles. One line of anti-bicycling medical discourse classed women's riding with other athletic activity as entailing objectionable amounts of exertion. Riding was seen as a threat to gender roles: while it would be safe for healthy men to ride, it would be dangerous for women to expend so much of their strength on physical activity. The other line of the argument against women's cycling framed its objections in terms of sexual health. Critics' concern focused on the supposedly problematic effects of bicycle saddles: as a result of the angle of the saddle, "the bicycle teaches masturbation in women and girls."[21]

The whole question of riding astride anything was problematic for women to begin with. Discussing children's toys, Karin Calvert points out that "one of the cardinal rules of child-rearing in the nineteenth century . . . forbade girls to sit straddling any object. Parents were advised not to let their daughters ride hobby horses, rocking horses, bicycles, tricycles, or even seesaws. . . . The position, parents believed, threatened the sexual innocence of their young daughters." The prohibition against girls using toys with straddle seats was part of the larger picture of rigidly separated spheres of play for girls and boys. "Playing the wrong games or using the wrong toys could prematurely awaken sexual feelings in children and destroy their natural purity," Calvert notes.[22] In other words, such play was a form of deviance that threatened both sexual innocence and gender definition. Not surprisingly, one physician called riding astride "too mannish to be proper for a woman."[23]

And yet with the advent of the safety bicycle and its female ridership, straddling could no longer be avoided, and attention focused on details of its troubling implications. In an outpouring of numerous articles in medical journals, physicians went into extensive and virtually prurient detail as to ways the bicycle saddle might produce sexual stimulation:

> The saddle can be tilted in every bicycle as desired. . . . In this way a girl . . .
> could, by carrying the front peak or pommel high, or by relaxing the stretched
> leather in order to let it form a deep, hammock-like concavity which would fit
> itself snugly over the entire vulva and reach up in front, bring about constant fric-
> tion over the clitoris and labia. This pressure would be much increased by stoop-
> ing forward, and the warmth generated from vigorous exercise might further
> increase the feeling.

This physician reported the case of an "overwrought, emaciated girl of fifteen
'whose saddle was arranged so that the front pommel rode upward at an angle
of about 35 degrees, who stooped forward noticeably in riding, and whose
actions . . . strongly suggested . . . the indulgence of masturbation.'"[24]
Although the patient was evidently worn to a frazzle by her fevered indulgence, the imagery of this physician's first passage seems to reflect concern that
female masturbation is a kind of indolence or relinquishment of vigilance: the
leather is "relaxed"; the vulva rests in that signal article of Victorian leisure furniture, a hammock. The tell-tale riding stoop of the second passage, however,
raises a different issue: it is the "scorching" position—that is, the bent-over-
the-handlebars posture adopted by speeders. In male riders it might be merely
criticized or mocked. But for women, fast riding was condemned; deviations
from upright decorousness and graceful riding were more serious, and a
woman's bicycle riding posture could be a significant measure of her propriety
and sexual innocence.

Another physician complained that "except when one rides slowly and
erect" the "whole body's weight . . . rests on the anterior half of the saddle."
Here, not only the saddle and its adjustment but also speed is at fault, and the
punishment for stepping out of line is pain and pathology:

> The moment speed is desired the body is bent forward in a characteristic curve
> and the body's weight is transmitted to the narrow anterior half of the saddle,
> with all the weight pressing on the perineal region. . . . If a saddle is properly
> adjusted for slow riding and in an unusual effort at speed or hill climbing, the
> body is thrown forward, causing the clothing to press against the clitoris, thereby
> eliciting and arousing feelings hitherto unknown and unrealized by the young
> maiden.

Painful and debilitating "granular erosion" or "polypoid growth" would result.[25]
Just as in the 1903 story "The Adventure of the Solitary Cyclist" Sherlock
Holmes's keen eye could read the bicycling of his client Miss Violet Smith from
"the slight roughening of the side of the sole caused by the friction of the edge
of the pedal,"[26] an educated eye might detect wantonness in any deviation from
an upright posture by a woman cyclist. Similarly, medical books had warned for
years that the signs that girls "are addicted to such a vice . . . [are] only too plain
to the physician"[27] and that the "habit" of masturbation left "its mark upon the
face so that those who are wise may know what the girl is doing."[28]

Thus, while bicycle ads might show men in a variety of riding positions,
women were shown only seated upright, as in figure 4-5. Moreover, perhaps in
line with the concern of medical books that *telling* girls of such a vice would be

in itself "dangerous," but would be unnecessary if other instructions were given properly,[29] guides for women cyclists didn't mention masturbation, but stressed instead the importance of sitting upright. Even bicycle enthusiasts Lina and Adelia Beard dwelt on the decorous upright posture: "Our eyes rest with delight upon the trimly clad, graceful rider who, sitting erect, seems almost to stand on her pedals as she moves along."[30] Manufacturers expressed this concern about posture verbally as well as in advertising graphics; the Rambler catalog proclaimed its woman's model's handlebars "so bent as to admit of an erect and graceful position, the secret of the well known fact that ladies when riding Ramblers present a more pleasing and attractive appearance than on any other machine."[31] Riding posture could be enforced, too, by the alignment of the bicycle: as figure 4-5 shows, women's handlebars were set several inches higher than men's, in accordance with instructions in one manual, and thus prevented women from "scorching."[32]

Whatever effect the rather appealing descriptions of bicycle saddles had on women's actual interest in riding is not clear. As we have already seen, however, women were an important market for the bicycle manufacturers that had sprung up in great numbers during this period, and opposition to their riding was an obstacle to sales. While the bicycle saddle may not strike us now as an ideal sex toy, manufacturers at that time took the medical discussion about women's bicycle saddles and masturbation literally and addressed it with the doggedly concrete and literal-minded solution of yet another product: a modified saddle that eliminated the point of contact with the genitals. Figure 4-6 shows ads for several crotchless cycle seats intended to circumvent possible masturbation. Ad copy for these seats typically warned of "injurious" or "harmful pressure exerted by other saddles" or declared their saddles "free of pressure

ALL RIDERS of REMINGTON BICYCLES enthusiastically praise these famous wheels. Many new features for '96 described in Catalogue, **free**.
REMINGTON ARMS CO., 313-315 BROADWAY, NEW YORK CITY.
BRANCHES:
New York, 59th St. and Grand Circle; Brooklyn. 533 Fulton St.; Boston, 162 Columbus Ave.; San Francisco, 418-420 Market St.

Figure 4-5   This Remington ad displays the greater range of riding postures allowed men, even including the man in the scorching position second from left. The differences in handlebar positions help to enforce the distinctions between men's and women's riding positions. With the exception of a few ads in which women are shown coasting with their feet up, women in bicycle ads are almost invariably in the upright posture shown here. (*Harper's Weekly*, May 16, 1896, p. 503)

against sensitive parts."[33] One ad delicately explained, for example, that it "is especially desirable for ladies, for it holds the rider like a chair, the entire weight being supported by the bones of the pelvis, which alone touch the saddle."[34] The connection between these euphemistic phrases and medical discussion of women's masturbation is clear from the medical articles from which some of the endorsements are drawn.

Of course the manufacturers' solution didn't get at what was really at issue: women's mobility and independence. The medicalized masturbation metaphor was a particularly compelling one because both the bicycling woman and the masturbating woman were out of male control, possibly doing damage to "the race." Conservatives feared that "masculinized" female cyclists would move out of traditional roles in other respects as well, making marriages outside of parental supervision and riding beyond the family and its control—issues soon to arise again in relation to the automobile. Certainly, as Marguerite Merington's and Frances Willard's celebrations of the bicycle as an escape from women's constricted sphere demonstrate, the possibility of escape was part of the *attraction* bicycling held for women. These issues, which were overshadowed by the focus on masturbation, were obviously too deep and complex to be affected by simply changing the bicycle saddle.

Benefiting the Race

While a medical journal might typically caution, "No woman should ride a bicycle without first consulting her medical man,"[35] medical arguments against bicycling were by and large restricted to the medical press and did not appear in the advertising-dependent general magazines; pro-bicycling medical arguments appeared in both. The pro-bicycling medical discourse was less likely to counter complaints about the saddle directly than it was to propose that bicycling might improve women's health by getting them into the fresh air. "Moderation," however, was crucial, and the physician was held up as a key figure to monitor and regulate the doses of riding, especially in anecdotes of women invalids helped by bicycling. One cycling book, for example, after first cautioning that any woman in less than perfect health should consult her physician before taking up cycling, reported the success of physician-supervised cycling for a semi-invalid. For this insomniac musician, her physician prescribed a series of cures: "It got to such a pass that the doctor stopped her music, shut up the piano, and forbade any hearing of music. This did no good. They then sent her to the mountains; this failed. The doctor prescribed a bicycle, but her mother would not consent, thinking something ought to be found less objectionable and just as powerful." When the doctor's variants of S. Weir Mitchell's famous rest cure fail, the doctor again insists on the bicycle and threatens to drop the case otherwise; the mother, here in the classic female role of conservative resistance to male modern science, agrees. The doctor's progressive insistence on the bicycle produces a cure and a grateful family.[36] Advertisers, too, tapped this discourse, in copy that assured readers that "bicycle riding is a boon

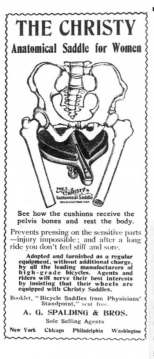

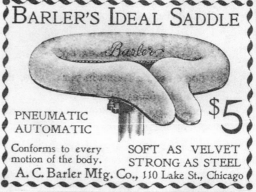

Figure 4-6 Ads for crotchless and modified bicycle seats were developed in response to concerns about women riding astride. Many of these ads referred to current medical discourse about riding. The Christy ad cites a medical booklet, while another ad trumpets, "You do not straddle the Duplex saddle."

Figure 4-6 Continued

119

. . . especially to ailing women"[37] (see figure 4-7), suggested that it put women "on the road to health" (figure 4-4), and explicitly proposed the bicycle as a medical prescription (see figure 4-8).

More crucially, the medically approved version of women's athleticism was often harnessed to a socially approved purpose: bicycling was commonly said to restore to health women whose invalidism and malingering made them unfit not only for musicianship but for motherhood, in part by strengthening the uterus.[38] Frances Willard, for example, exulted that "the physical development of humanity's mother-half would be wonderfully advanced by that universal introduction of the bicycle."[39] Advertisers asserted a direct link between motherhood and riding in ads such as the Columbia ad in figure 4-9.[40]

Redemption from invalidism into motherhood was a particular concern during a period of anxiety that white, nonimmigrant middle-class women were

Figure 4-7   This Rambler ad emphasizes the health benefits of the bicycle. (*The Century*, June 1894, ad p. 38)

An ounce of prevention is worth a pound of cure. Medical statistics show that on the average only one woman in a thousand is blest with perfect health. Is your wife an invalid? Are you constantly paying doctor's bills?

℞

Steel,
Leather,
Rubber,   *aa Q.S.*
Pint: *Safety Bicycle No. 1.*
Sig.   *Use frequently as directed.*

Physicians who will use this prescription will have no occasion to resort to Koch's lymph, cod-liver oil, etc.

Dr. C. E. RICHARDS.

The judicious use of a bicycle by a lady will work wonders in the improvement of her health. In constructing our

COLUMBIA LADIES' SAFETY

we have aimed to make it light, strong, and easy riding. The many testimonials we have received are the best evidence that we have succeeded in making a wheel perfectly satisfactory.

IT IS FITTED WITH

Cushion or Pneumatic Tires.

The pneumatic tire absorbs vibration and makes riding on any surface pleasant and agreeable. The handle-bar allows the rider to assume a natural and graceful position.

This machine and the pneumatic tire are fully warranted.

Apply for a Catalogue at the nearest Columbia Agency, or it will be sent by mail for two 2-cent stamps.

POPE MFG. CO.,

221 Columbus Ave., Boston.

| 12 WARREN STREET NEW YORK, N.Y. | 291 WABASH AVENUE, CHICAGO. | FACTORY, HARTFORD, CONN. |

When you write, please mention "The Cosmopolitan."

Figure 4-8 This Columbia ad emphasizes the health benefits available to women through bicycling, and points out that its handlebar design "allows the rider to assume a natural and graceful position," thereby preventing scorching. (*Cosmopolitan*, November 1891, ad p. 57)

having too few babies. (Given the high mid-1890s bicycle prices, these women constituted the female bicycle market.) Proponents of women's riding drew on this argument, too, to further support their advocacy. As Marguerite Merington put it, bicycling was "a pursuit that adds joy and vigor to the dowry of the race."[41] The word *race* here is used as it often appeared in the eugenics discourse of this period, where it ambiguously stood for both human and white native-born "race."[42] The Pope company's name for its bicycles—Columbia, the allegorical identity of the United States—and other companies' use of such names as the United States, the Patriot, the Charter Oak, the Eagle, and the Liberty ("America's Representative Bicycle"), already positioned bicycling as a particularly American, even patriotic, pursuit and made clear which group's fecundity would increase through moderate riding. Other manufacturers associated their bicycles with England, the source of what were seen as "real" Americans, and named their bicycles Imperial, Waverly, Worcester, Windsor, Warwick, Raleigh, Royal, and Richmond.[43]

Figure 4-9   In line with concerns linking physical fitness with reproduction, this Columbia ad links maternity and the bicycle. (*Harper's Weekly*, July 25, 1896.)

The Changing Magazine

The fears raised by women's bicycling were too deep and complex to be countered simply by changing the bicycle saddle or proposing pro-bicycling health arguments. And strategies used by individual advertisers to visually demonstrate women's graceful, feminine riding were insufficient as well. Instead, the threat posed by bicycling was defused through the advertising-dependent magazines addressed to middle-class readers.

As Christopher Wilson has noted, the ascendent editors of the ten-cent magazines of this period no longer waited in genteel fashion for stories to drop in over the transom, but actively solicited and commissioned topical, timely material.[44] Stories involving the new fad of bicycling appeared frequently, many of them in special bicycling issues. Tying magazine content to a fad lured advertisers; these special issues were filled both with bicycling stories and articles and with ads for bicycles, tires, cycling clothes, and of course saddles, to such an extent that bicycle ads constituted 10 percent of all national advertising in the 1890s.[45] (Magazines that kept their prices at twenty-five or thirty-five cents, such as *Scribner's* and *Harper's*, were also affected by the greater interest advertisers took in magazines.) Magazines were an attractive new advertising medium for a new product, but in addition magazines and bicycles shared other intimate connections, suggestive of the sympathy the new magazine publishers felt for the kinds of businesses likely to invest in advertising. Publisher S. S. McClure began his editorial career working for the Pope Manufacturing company as the editor of *The Wheelman* (later renamed *Outing*). In turn, *McClure's* magazine showed unusual helpfulness to Pope. In 1897, *McClure's* ran a four-part series under the heading "Great Business Enterprises: The Marvels of Bicycle Making," in which a *McClure's* writer toured factories of the Pope Manufacturing Company and reported in great detail the care taken in all phases of the bicycle-making process. Like the blind ads discussed in chapter 3, this series' layout was identical to that of the other editorial matter in the magazine; only a headnote at the beginning of the first piece in the series, and a small footnote at the end of the other pieces, told the reader that these glimpses "behind the scenes . . . [of] an advertisement" were paid for by Pope.[46]

Publishers and editors of the new magazines shared with their advertisers a common interest in the up-to-date world of commerce and industry and its new products—in this case, the bicycle. The monthly magazine's stake in timeliness encouraged the discussion and incorporation of the new advertised products in articles and stories. Doing what advertisers could not do for themselves, magazines acted for advertisers in the aggregate. The magazines' alignment and connections with industry encouraged favorable reflections on bicycles.

Because the fears women's bicycling raised were primarily social, fiction, with its articulation of social relationships, was better suited than medical articles or other coverage to take the sting out of those fears. It reconfigured the relationships the bicycle seemed to be changing and assigned new meanings to those changes. Fiction carried the burden of instructing readers in the complexities of the bicycle's social meaning, investing it with romance and glamour, and reassuring readers that riding would not disrupt the social order. Like the realist fiction in which Amy Kaplan finds a "strategy for imagining and managing the threats of social change," the formulaic stories examined here reassured readers that women on bicycles did not in fact threaten either the stability of the family or of just parental authority.[47]

Not all magazine writing about bicycles concerned courtship or defused or recontained the threat of women's mobility. Magazines of the 1890s featured bicycles in many roles: there were numerous and sometimes endless personal accounts of bicycle touring, such as *The Century*'s 1894 six-part series "Across Asia on a Bicycle," and stories chronicling fictional bicycle trips, such as humorist Jerome K. Jerome's serial "Three Men on Four Wheels" in the 1899 *Saturday Evening Post*. Bicycles also made incidental appearances in Christmas stories as the ideal present of a doting uncle.

Stories like Jerome's resemble the more straightforward ads: the bike works well or it doesn't; it goes fast or fails to; changing the tire is hard or easy. In them, the bicycle becomes the focus both as a mechanism in need of repair and as a narrative device for moving characters in and out of new scenery and groups of characters. As a prop or narrative convenience, the bicycle had many of the same attractions for story writers as it did for tourists: riders could ride alone or in pairs or groups; they could stop at will and visit familiar or unfamiliar places. Bicycling was, moreover, an eminently middle-class pursuit. The frequent appearance of bicycles in magazine fiction was a product of magazine editors' and publishers' identification with the interests of their advertisers, and it reflected middle-class authors', editors', and publishers' own participation in the amusements of their class, as well as the attractions of bicycling's use in narrative.

Bicycling also promoted new forms of heterosexual socializing that modified the forms of parental authority. One commentator at the time heralded the "new social laws" the bicycle was bringing into being: "Parents who will not allow their daughters to accompany young men to the theatre without chaperonage allow them to go bicycle-riding alone with young men. This is considered perfectly proper. It seems to be one phase of the good comradeship which is so strong a feature of the pastime."[48] Although this commentator's assertion that "every rider feels at liberty to accost or converse with every other rider" might be read as a threatening aspect of bicycle mobility, he points to the bicycle itself as enforcing "the uniformly quiet, orderly, and decorous conduct of the great army of wheelmen." While formal introductions were usually considered a necessity in the genteel and aspiring classes, meeting on bicycles seems to have escaped that requirement—perhaps because, at least through the mid-1890s, prices meant that riders could be assumed to be middle class. The bicycle functioned almost as another character: a mutual acquaintance who legitimately made the introduction. While one etiquette book suggested mounting a chaperone on a bicycle, this appears to have been rare both in the stories and in practice.[49] The stories instead asserted that the new, seemingly less controlled forms of heterosexual sociability enabled by the bicycle would produce at least as satisfactory a result as traditional courtship—and that those satisfactory results might even be caused by the bicycle.[50]

Bicycling Formulas

Stories that incorporated bicycles within a courtship plot (and courtship plots were ubiquitous in middle-class magazines), used formulas that defused the threat of women's bicycling. In courtship stories, bicycling frequently allowed middle-class young people to meet; usually they turned out to share both social class and social connections, and to be entirely suitable matches. In one such story, "Rosalind Awheel," by Flora Lincoln Comstock, which appeared in 1896 in *Godey's Magazine's* special bicycling issue, a girl resists her father's disapproval of bicycling.[51] The story addresses and defuses the possibility that bicycling might be masculinizing. Although Ethel dresses as a boy to escape home on her bicycle tour, she is soon spotted by a fellow bicyclist, a socially suitable brother of a girlfriend. Revealing that he has long been interested in her, he insists on seeing her home "safe under [her] father's roof." This story ultimately recaptures Ethel, but allows the reader to briefly enjoy with her the pleasures of transgression without permanent consequences. It closes with Ethel safely back in girl's clothes, her courtship with the right sort of fellow well under way.[52]

In another formulaic story, the desire for a bicycle enables romance and substitutes for longings toward more troublesome sorts of mobility. In Adelaide L. Rouse's courtship story "The Story of a Story," which appeared in *Munsey's* in 1896, Elizabeth tries to sell a story to a magazine to earn money for a bicycle. The magazine's editor and assistant editor mock the young woman's story to one another and reject it, but the assistant editor begins courting Elizabeth under the guise of teaching her to write, although he believes her incapable of it. The desire for the freedom of the bicycle and the power of writing and earning money are here both subsumed into romance. The story ends with the assistant editor's announcement to the editor that he and Elizabeth are buying a tandem bicycle—here both a sign of betrothal in an attractively companionable marriage and a kind of mobility specifically restricted, for the woman, to movement as a pair (the assistant editor already has his own bicycle). The bicycle story thus keeps an advertised item in the reader's eye while demonstrating its multiple uses and containing the threat posed by woman's mobility both literal and economic. Clearly, this young woman will not be riding off alone or, for that matter, making suspicious adjustments to her saddle. That threat has been undercut by watchful editorial control, both by *Munsey's* magazine and the betrothed editor.

The editors complain that Elizabeth's stories are "prose idyl[s]," set in the vague realm of romance, full of such sentences as, "The sun was shedding his last rays upon a lowly cot, embowered by trees, behind which flowed a rivulet."[53] Unlike such work, Rouse's own story is in line with the tenets of realism that her two editors espouse: it is full of the tangible life of the world. Its characters are understood through their relationship not only to one another but also to commodities like typewriters and bicycles. These signal our presence in an up-to-date world that is definitively different from life with lowly cots and rivulets. One result of this brand of realism, with its deployment of props, is that objects in the story are not just in the same two-dimensional space as the magazine's ads, but are in the same register as well: they are familiar to the reader from the ordinary middle-class world of commerce. And yet Rouse's story, with its neatly resolved courtship plot, is realistic only in contrast with the romantic excesses of Elizabeth's writing. To claim and even insist on that real-world affinity, Rouse's story depends both on this counterexample and on its assertion of commonality with the objects and advertising narratives in the magazine.

The events of Rouse's story are precipitated by a character's desire to buy a bicycle. The reader is invited to participate in the story's world of interested consumers. At the same time, through their presence in stories, the objects in the ads acquire significance and are endowed with a larger social meaning, including an association with courtship rather than with solitary, independent mobility. The bicycle links the courting characters both socially and mechanically. Elizabeth and her editor's mutual enjoyment of the bicycle points them toward new models of family life, in which, according to one commentator, "whole families ride together, carrying with them wherever they go the spirit of the family circle"; because "husband and wife are able to enjoy this together, the result is a new bond of union."[54] The bicycle thus facilitates new expectations of heterosexual sociability in courtship and marriage, a version of middle-class marriage in which a failed writer is a suitable helpmate to a rising editor. Advertising took up this urge as well, as in figure 4-10.

Rouse's story hints at endorsement of a new form of marriage in which the husband is less the absolute patriarch than the senior partner, still superior in knowledge and authority. Old-line patriarchal authority appears in other stories in the form of the father. When his authority is seemingly undercut by the daughter's bicycling, the daughter is shown to actually achieve a higher form of obedience to it. In one story, for example, an ailing young woman defies her father, the minister; she obeys the "prescription" of her brother, the doctor; and takes up cycling. She thereby gains renewed strength to apply to her father's housework.[55]

Similarly, Harry St. Maur's 1897 story "To Hymen on a Wheel" (the title alludes to the Greek god of marriage, not anatomy), emphasizes the old-fashioned unreasonableness of parental demands through its setting in a "wee village" in England, whose most "metropolitan" inhabitant is the postman who delivers mail on his bicycle. An obstinate, dialect-speaking father opposes his modern seventeen-year-old daughter Jenny's desire to marry Will: although he likes Will, her father wants Jenny to wait until she is thirty, her parents' age at marriage. The father secretly reads Will's letter, telling Jenny to meet him at noon to marry, and then thwarts her efforts to get to town alone. When she invents an errand, he follows her to Tim the postman's. Pointing out his bicycle, Jenny says

"Have you heard, father, as how girls ride them things?" indicating Tim's bike — "in trousers and breeches like men."

"The brazen things. No gal o' mine shall ever ride one in any kind o' way. It aint commonly decent."

Figure 4-10 The relatively late publication of this Rambler ad, in 1900, allowed for the inclusion of a child on a child-sized bicycle—a slightly later innovation. Having made the "choice of a life time" and committed themselves to a tandem, her parents ride alongside with a baby on the handlebars, carrying with them "the spirit of the family circle." The ad copy notes as well the shift from the male-only high wheelers of 1879 to the whole family's use of the 1900 bicycle. (*Harper's Weekly* April 14, 1900, p. 352)

Jenny gets on it, of course, and rides off to the registrar's, "flying down the road pedaling in great shape." In a scene not depicted, though it seems to have had something to do with a group of American tourists awheel who stayed at her father's inn, Jenny had learned to ride.[56]

Here, the bicycle *does* undermine parental authority. But we are carefully told how unreasonable that authority is, and that Jenny is obeying her father in almost every way—he approves of her choice of Will and even insists that her eventual marriage should be at the registrar's office. The father's opposition to women's bicycle riding becomes his final old-fashioned absurdity and affirms the righteousness of the daughter's move. Bicycling's liberating power is paired with an unambiguous situation. And, in an American magazine filled with ads for Columbia, Liberty, and United States bicycles, the cause of bicycling is positioned as a blow for American speed against old-world sluggishness.

The formula we saw earlier in the *Godey's Lady's Book* story "A Turn of the Wheel," in which the high-wheel rider is knocked from the world of male mobility in order to enter the female home and fall in love, is retooled for the safety bicycle in an 1893 story in the bicycling and outdoors magazine *Outing*. In John Seymour Wood's "A Dangerous Sidepath: A Story of the Wheel," Sam, a young, overworked lawyer, takes up riding after his doctor tells him he is "in danger of consumption." The story interrupts itself for a self-conscious testimonial: "Today he is sound as a new dollar, his eye bright, his lungs healthy. But how this sounds like some medical 'ad'!"[57] It is not an ad, the narrator explains, since he does not promote a particular model. He goes on to tell Sam's courtship story.

On a week-long bicycle trip with two male friends, Sam rides on a small town's smooth sidewalk until he is stopped by a pretty young woman. She turns out to be Nathalie, the cousin of one of Sam's companions, in town for the summer. Her family invites the three men to stay with them, but Sam, now enamored of Nathalie, declines. He walks past her house instead: "He saw a lady's wheel leaning against a pillar of the porch. Did *she* ride? Instantly he thought of her flying along under the elms, a vision of beauty on her wheel—for a pretty girl never looks so pretty as when wheeling." Unlike the young man on the highwheeler in Provence smitten by a statuette, Sam is transfixed by the vision of Nathalie in motion, and her motion is reformulated as yet another visual treat.

Finally the opportunity to talk arrives:

> "*You* are fond of riding?" asked Sam, astonished.
> "I do nothing else."

And they have even more in common: both like to stop and read and dream along the way. They joke that they scorch downhill only.

The three travelers set out for a moonlit ride with Nathalie and her brothers. The visual pleasure of watching her cycle excites Sam's admiration still further: Nathalie rides "to perfection, sitting very straight and running the wheel on a line." She daringly leads the way crossing a stream on a single plank. The rest follow safely, but Sam, distracted by love, falls in and breaks his leg. Back at her house for the requisite nursing, Nathalie contritely ministers to him. Following the inevitable engagement, the couple "took their wedding journey on the latest of [bicycles] . . . and this summer, Sam Selover is wheeling, on the sidewalks of Farley—a baby carriage."[58]

Although Nathalie is an active participant here and precipitates Sam's fall into love, the stream, and her household, her role remains little different from that of the Provençal statuette. On the other hand, life beyond the boundaries of the formula is somewhat shifted: the two will continue to venture into the world on their bicycles, and it is not only the mother of the couple (her uterus no doubt strengthened by riding) who exchanges her bicycle for a baby carriage. The use of this formula, however, reassures the reader that women in motion on bicycles (as opposed to standing still in doorways) do not excessively disrupt convention, even literary convention.[59]

Writing against the Formula

That the formula of the injured man rescued into marriage came to be considered something of a laughable cliche is suggested by an 1897 parody of this pattern, Charles Day's "Willing Wheeler's Wheeling: A Bicycle Story," in which a man's attempt to fake an injury and thus woo Miss Finlayson, "*a pretty girl on a bicycle!,*" while being nursed goes awry. His plan has been fed by reading too many romances.[60]

Writers like Merington and Willard showed women reveling in the freedom afforded by bicycling. But because letters to the editor rarely appeared in this period's magazines, there is little direct evidence of readers' responses to the formulaic stories. The threat the formulaic stories counter had been articulated in terms of both women's sexuality and independence. An explicit awareness of and response to these formulas does appear in two stories that subvert, rather than simply parody, these formulas. Both were published in magazines that did not depend on bicycle ads. Here the threat posed by women's cycling is suddenly visible.

The new athleticism for women and girls increasingly permitted and promoted in this period promised the new woman that she could leave behind the specter of invalidism, that is, her body's own potential for invalidism. And in Kate Chopin's 1895 story "The Unexpected," the bicycle facilitates escape from an invalid body—externalized as someone else's invalid body. Rather than recontaining the threat to propriety and the established order implicit in women's bicycling, "The Unexpected" vividly dramatizes and celebrates it.

Randall and Dorothea are a passionate engaged couple who are separated when Randall contracts a wasting disease. Dorothea ardently anticipates his visit on his way south to recover, but finds herself appalled and repelled by his transformation: "This was not . . . the man she loved and had promised to marry. . . . The lips with which he had kissed her so hungrily, and with which he was kissing her now, were dry and parched, and his breath was feverish and tainted."[61] Randall still asserts a hold on her, however: "All the strength of his body had concentrated in the clasp—the grasp with which he clung to her hand." He wants them to marry immediately so that she can come along and nurse him or, alternatively, have his money when he dies. Interrupted by a coughing fit, he is helped away by his attendant before she can answer. We feel her revulsion and understand that convention is about to close in on her: "She realized that there would soon be people appearing whom she would be forced to face and speak to." The expected outcome seems to be that she will do the proper and dutiful thing: set aside her own reactions, marry him, and go with him to nurse him. But the story takes an unexpected turn: "Fifteen minutes later Dorothea had changed her house gown and had mounted her 'wheel,' and was fleeing as if Death himself pursued her." She rides far:

> She was alone with nature; her pulses beating in unison with its sensuous throb. . . . She had never spoken a word after bidding him good-bye; but now she seemed disposed to make confidants of the tremulous leaves, or the crawling and hopping insects, or the big sky into which she was staring.
>   "Never!" she whispered, "not for all his thousands. Never! never! not for millions!"[62]

Bicycling allows Dorothea to escape duty and self-sacrifice, as well as strictly material considerations. Pressed to answer yes or no, she takes to her wheel and escapes the deathly embrace of marriage. She rides off, pulses beating, into a sensuous, edenic solitude in which she can refuse the marriage. In this story, bicycling arrives in high gear, brakes off, as a threat to the social order. It allows a woman not only the right, which might elsewhere be developed in abstract discussion, but also the wordless *power* to move beyond duty and affirm her physicality—here, set in opposition to the grasping deadly power of diseased marriage. It moves her far beyond the familiar trope of the woman as nurturing caretaker to a conclusion that takes the formula in an unexpected direction.

It is significant that "The Unexpected," in which bicycling figures so unsettlingly, was published in *Vogue*, which at the time was less a fashion magazine than a society weekly and often printed material that twitted conventional bourgeois views. Emily Toth's discussion of Chopin's publishing history suggests that *Vogue* took stories from Chopin that the more conventional *Century*, *Harper's*, and *Scribner's* would not accept.[63] Its circulation was small and, moreover, it did not depend heavily on advertising for its revenues; as one magazine historian puts it, "its advertising business had never been properly developed."[64] The ads in *Vogue* were mainly for carriage trade retailers—silversmiths, Louis Sherry, and carriage makers themselves—rather than the ads for manufacturers like bicycle makers, which appeared in the pages of *Munsey's*, *McClure's*, *Godey's*, and other magazines that depended more heavily on ads; or in *Outing*, which was subsidized by a bicycle manufacturer. As we have seen, these magazines printed fiction that defused this apparent threat, in contrast with *Vogue's* willingness to flaunt the license accessible to women through bicycle riding.

The sense that bicycle ads and the story are in the same register, however, is fleetingly present. Chopin's story provides a reading of what we saw in figures 4-3 and 4-4, a common image in bicycle magazine ads and catalogs—that of a woman riding off into the countryside. "The Unexpected" rereads and rearticulates the appeal of such images, and plays out the fantasy of freedom that advertising graphics invite the female viewer to enter; Chopin, however, does not make it safe for those who might disapprove of such independence by reassuring the reader that Dorothea's costume covers her toes. We do not know whether Dorothea sits very straight and rides "to perfection," like Nathalie. Dorothea is not offered as the object of spectatorship, an image of appealingly decorative riding, "a pretty girl on a bicycle," but as someone using the available technology to make a break for freedom. Chopin thus wrenched the ad imagery from its lulling context and reread both it and the threat of women's riding as a heartening escape from constraint.

Chopin's anti-courtship story, in which a courtship is unraveled and only the woman rides off into solitary pleasure on her bicycle, drew on and celebrated the same metaphoric connection between women's independence and sexuality on the loose that the medical discourse about bicycling and masturbation found so alarming. An 1896 Willa Cather story, "Tommy the Unsentimental," played out the bicycling woman's threat to gender definition.[65] Unlike the cross-dressing Ethel of "Rosalind Awheel," whose excursion leads to both reconfinement in her father's home and a suitable match, and unlike a character in another 1896 story, whose cross-dressed fast ride leads to embarrassment,[66] the ambiguously gendered Theodosia, called Tommy, uses her riding prowess to rescue Jay Ellington Harper, the foppish, ineffectual young cashier from the East to whom she is ostensibly engaged. Tommy is the no-nonsense daughter of a banker in the West, where "people rather expect some business ability in a girl . . . and they respect it immensely." With her good sense and acumen she repeatedly bails Jay out by doing his work for him, and the attentions he shows Tommy in exchange are attractive to her. And yet Jay clearly fails to properly appreciate her. The people who do are the Old Boys, her father's old business friends. Tommy reciprocates their feeling for her in her Western way, by playing billiards with them, mixing their cocktails, and taking one herself occasionally. Jay, of course, disapproves of such acts, and the Western Old Boys disapprove of Jay.

The story layers an alternative model of gender onto the conventional one. Here, whether people are male or female is sometimes less central to either their sexual preference or gender roles than whether they are Eastern or Western. The Old Boys, for example, have "rather taken her mother's place."

When Jay's father buys him a bank of his own twenty-five miles away, Tommy often bicycles over "to straighten out the young man's business for him." The Old Boys are concerned when Tommy goes East to school for a year: "They did not like to see her gravitating toward the East; it was a sign of weakening, they said, and showed an inclination to experiment with another kind of life, Jay Ellington Harper's kind." Nonetheless, she does well at school with what counts for the folks back home—athletics—and returns, acknowledging that Easterners are not her sort. But she brings back with her a prototypical Easterner, "a girl she had grown fond of at school, a dainty white, languid bit of a thing, who used violet perfumes and carried a sunshade." Miss Jessica makes an impression on Jay, who evidently senses a kindred spirit. While Cather leaves ambiguous the question of whether Tommy will be losing Jessica to Jay, or Jay to Jessica if they get together, Tommy's heroic bicycle ride at the climax of the story makes clear that the two Easterners are fit only for one another, and that neither deserves Tommy.

When Jay telegraphs that he needs help by noon against a run on his bank, Tommy decides to bring him money to pay off his depositors, though the only way to do so is a difficult bicycle trip. Miss Jessica, seeking a grand gesture to impress Jay, comes too. But when Tommy refuses to stop to rest and drink water and instead "bent over her handle bars, . . . it flashed upon Miss Jessica that Tommy was not only very unkind, but that she sat very badly on her wheel and looked aggressively masculine and professional when she bent her shoulders and pumped like that."[67] Miss Jessica drops out half-way along, asking Tommy to tell Jay that she would do anything to save him.

Miss Jessica sees in Tommy precisely the sort of riding that manuals warn women to avoid, and which, we have seen, is associated with saddle mastur-bation. Tommy's scorching posture and lack of moderation, however, bring results in an extraordinary riding feat. The 1896 record for men on racing bicy-cles on a paved flat track was twenty miles in forty-five minutes. Tommy rides twenty-five miles on a rough unpaved uphill road in the hot sun, not on a rac-ing bike, and carrying a heavy canvas bag, in seventy-five minutes.

She arrives in the nick of time, moneybag in hand, scolds Jay for his poor business practices, and tells him where he can find Miss Jessica: "I left her all bunched up by the road like a little white rabbit. . . . I'll tend bank; you'd bet-ter get your wheel and go and look her up and comfort her. And as soon as it is convenient, Jay, I wish you'd marry her and be done with it. I want to get this thing off my mind." Jay is shocked; he thought he was engaged to Tommy. Mas-culinity once again becomes an operative term as he thanks Tommy for what she's done for him: "I didn't believe any woman could be at once so kind and clever. You almost made a man of me." "Well I certainly didn't succeed," she replies, and sends him along.[68]

Tommy has played for a time the role familiar from the formulaic stories of helping out the immobilized man, here immobilized by his own shortcom-ings instead of by an accident. But she is not interested in reducing her sphere and bringing him into it, in entering a sphere as limited as his own, or in bring-ing both into the "spirit of the family circle." A life of fast riding in the larger sphere of action and Westernness is clearly superior to either Jay's world of incompetence or Miss Jessica's adjoining world of decorum. Concern with decorum may promise to make its practitioners fine ad illustrations, but actu-ally leaves one of them bunched up by the side of the road "like a little white rabbit."

Although Cather and the story approve of Tommy and of behavior classi-fied as "Western," Cather's portrayal of Tommy's riding clearly links leaning over, scorching, and immoderation in riding with "masculine" traits and atti-tudes. And although Tommy has used her bicycle to visit her ostensible fiancé, bicycling here is not recuperated or promoted as an advantageous new form of heterosexual socializing that leads to a "new bond of union." Rather it becomes a means for a woman to more fully encompass and investigate the world, a test dividing the autonomous woman from the fops and the rabbits.[69]

Cather's story was published in *The Home Monthly*, intended as a genteel and moral entertainment; as Cather's biographer Sharon O'Brien notes, Tommy's cocktail-mixing abilities were hardly the skills the publisher wished to recommend to his teetotaling readers.[70] But the story appeared in the first issue that Cather edited, for which she wrote half the copy to make up for the absence of a file of manuscripts on hand. Under the circumstances she may have felt licensed to use work she had already written, or to please herself by publishing a story that subverted the conventions of bicycling stories, as she does here. Perhaps it helped that *The Home Monthly*'s ads were mostly for local merchants, not national manufacturers.

Summary

The nationally distributed middle-class magazines took on the project of defusing a threatening commodity—the bicycle—and making it salable by asserting scenarios in which it upheld and renewed the traditional social order. Within these scenarios, played out repeatedly within formulaic fiction, bicycles no longer threatened to erode gender distinctions or to make riders loose women.

The ad-dependent magazines' makeover of women's bicycling in a more decorous image may have been appreciated by women who wanted to ride, as is suggested in accounts such as that of Annie Nathan Meyer, a founder of Barnard College. Meyer took up bicycling early and touted her approach to it as an example of her "shrewd theory that to put any radical scheme across, it must be done in the most conservative manner possible." In contrast to Willard's and others' advocacy of the bicycle costume as a first step toward dress reform, Meyer adopted the protective coloration of a "well tailored suit . . . and modest hat" so that "there was nothing about me to indicate that I was not a genteel conservative, though I was engaged in an exceptional and pioneering act."[71]

Defusing the bicycle's threatening aspects through attractive images in ads and through formulaic fiction in mainstream magazines perhaps provided similar protective coloration and made it easier and more acceptable for women to take up bicycling. In fact, women made up an estimated quarter to a third of the bicycle market by the end of the decade. The magazines recast or read out of the discourse the possibility that bicycling might send women riding off alone, outside of or away from marriage. The stories instead suggested that the bicycle was simply an aid to creating an improved model of heterosexual marriage.

Roads branch out in many directions from women's bicycling in the 1890s. The marketing of bicycles appears in retrospect as a test drive for the subsequent marketing of the automobile, where again we see differentiated men's and women's versions, rapid deployment in fiction and social experience as an aid to courtship, and a mixed impact on women's lives.[72] Bicycling undoubtedly prompted dress reform, as it helped make lighter, less restrictive clothing acceptable. It made new forms of women's athleticism acceptable as well, and pressed toward greater freedom of travel for women.

The "bicycle craze" ended around 1898.[73] Prices dropped drastically— from $150 in the early 1890s to $10 by the end—because of production improvements and, arguably, because the well-to-do market was saturated. Women and men continued to ride, but bicycles were no longer a genteel novelty, and manufacturers no longer found it profitable to advertise heavily in middle-class magazines. Editorial content followed advertising, and stories and articles featuring bicycles faded out of magazines about that time.

Working together within the larger framework of the magazine, advertising and fiction made a seemingly threatening new product attractive to potential users. While ads could address a specific manifestation of the threat by promoting a new product like the "hygienic" saddle, the larger issues raised by women's increased mobility couldn't be headed off as easily. Magazine stories took on those issues by rewriting the product's apparent threat to traditional roles. They subsumed the potential conflict within a discourse of consumption. Soon after, automobiles began appearing in magazine stories—but that's another story.

5

Rewriting Mrs. Consumer: Class, Gender, and Consumption

Lately there has been a great deal of fretful, impatient, womanly writing, about the degrading, depressing influence of household work; and it has been urged that it is better for wives and mothers to write or sew, or do any kind of mental work, in order to make money to relieve themselves of the duties of cooking and nursing.

Amelia Barr, *Ladies' Home Journal*, 1894[1]

Edward Bok, the editor of *Ladies' Home Journal* from 1889 through its steady rise to a circulation of 2 million in 1919, was sensitive to the toll taken by restrictions on women. In two paradigmatic episodes in his autobiography, Bok noted that such restrictions confined and immobilized women. Where another writer might have expressed indignation on their behalf, or called to change the situation, Bok responded by seizing opportunity: a women confined was a captive consumer.

In the first of these scenes, Bok, writing of himself in the third person, explains that the horse-car line near his childhood home ran to Coney Island: "Just around the corner where Edward lived the cars stopped to water the horses on their long haul. The boy noticed that the men jumped from the open cars in summer, ran into the cigar store . . . and got a drink of water from the ice-cooler placed near the door. But that was not so easily possible for the women, and they . . . were forced to take the long ride without a drink." Bok recognizes a problem when he sees it. "Here was an opening, and Edward decided to fill it. He bought a shining new pail, screwed three hooks on the

135

edge from which he hung three clean shimmering glasses, and one Saturday afternoon when a car stopped the boy leaped on, tactfully asked the conductor if he did not want a drink, and then proceeded to sell his water, cooled with ice, at a cent a glass to the passengers."[2] He makes a handsome profit, first on Saturday afternoons, and then after morning Sunday school, "he did a further Sabbath service for the rest of the day by refreshing tired mothers and thirsty children on the Coney Island cars." When other boys move in on his idea, he upgrades to lemonade. Furthermore, Bok calls attention to the cleanliness of the service he provides, with its shining pail and shimmering glasses, supplying customers with an assurance of purity. And his profitable sales are a pious act, not only no violation of Sunday propriety, but a Sabbath service.

In the second episode, as an eighteen-year-old covering theater news for the *Brooklyn Eagle*, he "noticed the restlessness of the women in the audience between the acts. In those days it was, even more than at present, the custom for the men to go out between the acts, leaving the women alone." Women restless? Bok identifies an opportunity: "Edward looked at the programme in his hands. It was a large eleven-by-nine sheet, four pages, badly printed, with nothing in it save the cast, a few advertisements, and an announcement of some coming attraction."[3] A smaller, easier to handle, more attractive program might do the trick. Bok goes home, makes up a dummy, offers the eager theater management the program for free, solicits advertising, and soon expands the enterprise to cover captive audiences in all the city's theaters. Bok learned from and eventually repeated his theater program experience on a larger scale with *Ladies' Home Journal*: immobile women made the best customers.

Bok was a powerful figure in the *Ladies' Home Journal*; the *Journal* in turn influenced other women's magazines and shaped what advertisers could expect from a women's magazine. Bok's stance, promoting women's relationship to consumption, and ultimately opposing women's autonomy and ability to earn, was taken up by other middle-class women's magazines and in retrospect seems an almost inevitable strategy for them.[4] And yet not all commercial women's magazines of the 1890s to 1910s took this tack, although all wanted their readers to buy goods from the advertisers on whom their revenues depended.

The proliferation of nationally distributed brand-named goods in the 1880s and 1890s coincided with, and fed the growth of, the ad-dependent magazines; many magazine ads were for household goods, and ad-dependent magazines therefore had an interest in encouraging consumership. This encouragement took different forms, according to the presumed class and living situation of the reader addressed by different magazines. Rural and poorer women were urged to find their way into middle-class consumption practices, by buying more ready-made goods to replace their own labor, and so stories addressed to them suggested ways to earn money they could spend on advertised goods.

On the other hand, stories in magazines like the *Ladies' Home Journal*, addressed to women who were already presumed to be middle-class shoppers, assumed that such women already had money to spend and discouraged such work in favor of life as a married consumer. Fiction in these magazines indirectly made shopping a theme. Female authorship as a topic in stories in both types of magazines became an emblematic form of work for pay. The magazines' attitudes toward it correlated with the editors' and publishers' assumptions about their readers' sources of money for advertised goods.

Promoting Consumption by Encouraging Earning: The Mail-Order Magazine

Although 1893 saw the appearance of a new category of middle-class, advertising-supported magazines, soon to achieve high circulations and aimed largely at urban readers, another category of periodicals had preceded such magazines as *Munsey's* and *Cosmopolitan* in the low-price, high-circulation strategy. These were the cheap mail-order monthlies, such as *Ladies' World*, *Comfort*, and *People's Literary Companion*.[5] Usually nearly tabloid in size, addressed to rural and poorer subscribers, these magazines mostly ran ads that solicited mail orders, rather than advertising the nationally distributed goods that predominated in middle-class magazine ads;[6] some of them offered mail-order goods themselves, and the line between buying goods from a mail-order magazine and receiving a premium with a subscription could be thin.[7]

Advertisers faced a special problem in addressing the reader of such periodicals: she was likely not to have cash. She was also more likely than the middle-class woman to make rather than purchase goods. Early 1890s *Ladies' World* columns reflected this orientation with recipes for home manufacture of such items as skin lotions and beef extract—items already commercially produced and heavily advertised in the middle-class magazines.

Cash brought in through such traditional women's market activities as butter, egg, and poultry sales might already be allocated for ongoing expenses and not available to buy the new advertised goods. Individual advertisers recognized the readers' situation, with ads for poultry incubators and by offering goods such as china as a premium with staples like coffee. Many ads offered entrepreneurial opportunities; they often both touted a product such as flower seeds and sought agents to sell it.[8]

While both articles and stories in these magazines showed readers ways to earn money, the stories demonstrated the social consequences of earning and spending. The low-priced (twenty-five to fifty cents a year) monthly *Ladies' World*, edited in the 1890s by Frances E. Fryatt, offers some examples.[9] *Ladies' World* stories repeatedly suggested that women could earn money by starting a business or by taking on for money some new task or variant of a task they already performed, and that doing so would not only enable but also legitimate consumption. The similar plots of several stories reflected the same suspicions about the source of goods and concern with how commodities had arrived in a household.

"One Woman's Way" (1890) by Velma Caldwell-Melville begins with the narrator visiting "John's wife," while John's mother complains of the wife's extravagance in purchasing a $35 bedroom set. After the mother-in-law leaves, the wife explains that she *earned* all the money for the set. The narrator is surprised: she knows how busy a farm woman must be, caring for a husband, two babies, and two hired men. Money from the milk, butter, and poultry she sees to is already earmarked for necessities. The bedroom set? John let her use land to grow berries on, her father gave her plants, and she bartered her sewing for a neighbor's labor. She has used the money from each year's berry sales to make the home—her workplace—more comfortable, for example, by buying rocking chairs. Her up-to-date economic orientation is contrasted with her mother-in-law's outmoded one: "John's mother boasts that she has neither earned nor wasted a cent since she was married."[10]

To neither earn nor waste, however, is no longer a virtue; the contrasting condition is not of having nice things but, according to the stories, greater efficiency and even economy. In the 1904 *Ladies' World* serial "The Rebellion of Reuel's Wife" by Adella F. Veazie, Mazie's hatred of housework actually leads her family to greater ease. Beginning with a flowerbed for her own pleasure, she starts a flower seed business and soon branches out into landscaping, bringing in money to support the family when her husband is disabled. *Her* mother-in-law objects even to the first flowerbed and counters Mazie's defense of it as a respite from housekeeping by championing old-fashioned household production:

> "If I got any chance to set down an' rest, I always had my knittin' work handy so as to keep busy at something *useful*. Why, I've always knit every stockin' an' mitten that Nathan ever wore since I married him, and all my own too, except thin summer ones."
>
> Mazie gave a little gasp of dismay. "Why surely you wouldn't expect me to knit stockings now, when they can be bought for less than the yarn would cost," she exclaimed.[11]

For the mother-in-law, visible industry is a moral virtue; even at rest she produces goods, and her labor is visible on the hands and feet of her husband. But for Mazie, the stocking need not be a direct representation of the labor and care that went into it. Mazie's understanding of new economic possibilities allows her to see that cash lets her replace her own inefficient labor with manufactured goods; the more she earns, the more she is entitled to replace her labor.[12] The same idea appears in M. Vaughn's 1891 "John's Wife," elaborating the pattern of the other John's wife of the berry story, "One Woman's Way."[13] Here again, a woman appears to violate the propriety of thrift or moderate consumption. But the truth reveals her as virtuous, shows that her actions have been misconstrued, and lets her teach a lesson to those who have speculated about her.[14]

Jerusha and her mother believe cousin John's wife, Claribel, spends beyond his means. Claribel has hanging lamps and china dishes, and, in a farm woman's vision of luxury, she hires people to wash, iron, clean, and bake, and has "all their best clothes made, to say nothin' of *buyin'* them." She also indulges in charity, spending money to help the town's poor and invalids, and buys children's books and pretty "Scripture text cards" to increase Sunday school attendance.

Jerusha and her mother righteously confront John and Claribel. Claribel explains: she formerly wrote for a few magazines and had planned to give it up when she married:

> But loving the work and being impressed by the poverty in and about the village, also the lack of interest on the part of the children in school, Sunday-school, or in fact, anything good, she had resolved to again take up her pen, and *by careful management she could put out a part of her work and make much more than she could save by trying to do it all herself.*
>
> "I have bought some good books and a few pretty things for my house, although I have gotten nothing new to wear." (emphasis added)

Even this explanation of her work as selflessness, hiring out the housework as a better use of her time since her authorial labor has market value, isn't sufficient: she must absolve her earnings of the taint of producing pleasure or luxury for herself in the form of new clothes:

> "Nothing new!" spited Jerusha. "What did you have on in town last week? And what did you wear last Sunday?"
>
> "In town, I wore my lavender wedding-dress, colored a dark blue; and on Sunday, a white Flemish tricot (the dress I graduated in), colored black," was the quiet answer. "I flatter myself that I have two very respectable suits from them."

Even as she divests herself of some household tasks, she thus demonstrates her handiness as a housekeeper. And she had avoided telling her in-laws about her writing because she believed they'd think her "silly to suppose I could write anything worth publishing; what I wrote seemed so insignificant to me that I did not want anyone I knew to read the wretched 'yarns' I spun."

Although the narrator assures us that Claribel did not go on to be a great writer, her story has taught others. From it, Jerusha "learned not to be curious, not to be suspicious, not to be envious, and not to 'jump at conclusions' when they were based only on circumstantial evidence, and above all not to whisper words that might cause any to be misjudged."

Women's authorship here is strictly a money-making proposition—very much in line with the cash-raising schemes presented in other stories. Not only is Claribel's writing said not to be artistic, but Claribel's moral effect doesn't come through her writing. Instead, it is channeled through her power as a consumer; she does good through what her earnings buy for the community. Paying people to do her housework is finally sanctioned because she has taken up the moral housekeeping of the community.

Stories about married women earning money often pointed to tension around the gender-appropriateness of female money-making or of avoiding household tasks. These tensions were sometimes resolved by disabling and feminizing the husband, at least briefly, as in "The Rebellion of Reuel's Wife," where Reuel's disabling accident lets him see how necessary Mazie's work is. The need for female-generated income is thereby accentuated, and the husband thereby comes to see the value of the wife's work.

In "John's Wife" there is less tension around the gender-appropriateness of Claribel earning money than around the class-appropriateness of her spending: she might disrupt the careful balance between her husband's earning and her consuming. As an outsider (she is not a farm woman by birth), she may be seen as importing unfamiliar urban patterns of buying rather than producing—here, by buying clothes ready-made instead of making them herself. Her consumption, however, is framed within an ethic of thrift. The stories attempt to make the two compatible: John's wife in "One Woman's Way" buys a bedroom suite, and Mazie buys economical ready-made stockings with the same sort of enterprise.

Claribel not only seems to buy clothes, but to do so in extraordinary quantity, although that turns out to be an illusion wrought by her use of the modern, mass-produced commodity, clothing dye. In the element of the story most accessible to imitation by readers, industrial magic makes a dress unrecognizable even to relatives' prying and technically knowledgeable eyes. At this point "John's Wife" almost seems like an ad for clothing dye. So it's not surprising that seven years later the Diamond Dye company, a constant advertiser in *Ladies' World*, adopted virtually the same structure and some of the same details in its 1898 advertising booklet, *Cousin John's Extravagant Wife*, by Emily Hayes.[15] Here, yet another John seems to require his relatives' voluble advice. Annie is an apparent spendthrift, and his relatives worry that she's clearly living beyond John's means, wearing new dresses often, acquiring three new feather plumes since they've known her, and displaying various luxuries for the house. The relatives confront John and his Annie. Following the familiar structure of revelation, the prying relatives "were met with the usual cordial welcome":

> "John has told me," continued Annie, "of your kind interest in his behalf, and mine also, I truly believe, and suggests that I give you a little history of my extravagant purchases. This," she said, taking a little box from the table beside her, "contains the secret of my new feathers and hats and dresses, my new brown jacket and knitted capes and shawls, the new curtains and table-covers, the new afghan and Turkish rug, my olive dress, the bronze vases, the clover-leaf table, the hall stand, and many other little things which help to make our home attractive."
>
> She opened the box and took out a quantity of little envelopes.
>
> "Diamond Dyes!" exclaimed Aunt Margaret in surprise, while Aunt Maria for once was speechless from astonishment.[16]

The secret is revealed: she dyes her old plumes and clothes, or takes the clothes apart and dyes the wool and makes it up into something else. The bronze vases are plaster, painted with Diamond paint, the table is cheap wood stained with Diamond cherry stain, and the carpet is made of scraps dyed and reworked. In contrast to the story "John's Wife," each project Annie describes is extremely labor intensive, but she has invested little money for materials.

The story line allows for enumerating the items and their attractiveness several times, and of course the fact that the relatives thought Annie was extravagant means they thought the items were new: dyeing didn't occur to them. Within the story's melodramatic structure, the box Annie brings out should contain a legacy; within the pattern of the other stories we've seen, it could contain berry-raising receipts, or payment for stories or articles. Instead, what it contains is instantly recognizable, but how she used the dye to get the results she did is not, since she has remade the objects as well as dyeing them. The structure allows the reader first to relish the luxuriousness of the goods, then, through the revelation, to find them affordable, and through the explanation imagine herself making them. The explanation goes far beyond the exposition of berry-raising schemes as it embeds detailed instructions in the narrative:

> "Then my little shawl, which you think so pretty, is made from the crocheted cape my sister gave me at Christmas, two years ago. It was pale blue, you remember, and I had washed it several times; so I raveled that, wound the wool into a rather short skein, and tied it in three places, then washed and boiled in soap and water, according to directions, and while it was quite damp dipped it in the dye, and it did take a lovely shade!"[17]

The story suggests that time and labor were plentiful for rural women, while materials were scarce. Its constant envious or loving description of the goods establishes that consumption is bounded and watched: all eyes notice what is worn, remember it in detail, and know when it was acquired. The reader is invited to participate in an inner surveillance as well, enumerating her own goods in the expectation that others might accuse her of extravagance. And yet this enumerating is also an invitation to feel dissatisfied, to review each garment and think of how it might be improved by Diamond Dyes. The appeal of the story is in tricking surveillance, in having the magic box at hand with its explanation, but the appeal depends as well on being aware of that watching eye to which one must answer.

The report of extravagance in the advertising booklet, as well as in the stories, is thus transformed into one of thrift, economy, and handiness. Interestingly, neither Annie nor Claribel is accused of wasting time by messing around with dyes and cutting up and remaking her clothes. A woman's desire to dress well is seen as legitimate, as long as money is not spent on it.

All these stories assume in common that people closely watch over one another's purchases and possessions and know one another's income. They assume, too, that a woman's spending habits might bring ruin on her family and that such surveillance is therefore justified. The plot twist in two of the stories, however, embodies a rebuke to the watchers: they turn out to have misunderstood the situation and are embarrassed by their wrong guess. So the reader learns that it is perhaps best not to chide others for their spending. Each story points out that there are legitimate ways to seem to live beyond one's husband's income, legitimate ways to acquire more goods or more fashionable goods—and that others may have found such means.[18] In this case, ads and stories not only have structures, techniques, and concerns in common, but ads borrow from stories to the point of plagiarism.

In *Ladies' World*, writing is both a way to earn money and a source of self-reliance and independence. In Bertha Ashton's children's story, "Aunt Crawford's Wise Will; or Perseverance Conquers All Things" (1892), Aunt Desire Crawford threatens to cut her twenty-two-year-old namesake niece out of her will for her indolence, but will leave the money on the condition that "ten years from now you have made a name for yourself." As a child, young Desire wrote "short but bright stories," but expecting the inheritance, she has grown lazy. Angry at the condition set by her aunt, she turns to the alternative:

> "It's hard work to write steadily, but I can, and yet all this trouble would be saved if I married Philip Astor."
>
> "You do not like him; you would not be happy with him," her better self reasoned. "But his money, think of that. Work is irksome. An easy time you would have as his wife," something else whispered to her.

Her better self wins the argument, and she resolves "to please her aunt and be a better woman" by writing children's stories. She achieves fame for her writing, "calculated to make the children happier and more contented with their lot"; at the same time, evidently without contradiction, she "never forgot to instill in her stories the wish to be independent." The lot with which children reading *Ladies' World* were to be made contented was unlikely to include the option of marrying a rich husband like Philip Astor; it might well include the need either to support themselves independently or to continue to earn money after marriage. The story supplies detailed encouragement for doing so and concrete information about what to expect, showing young Desire persisting in her writing past the first few rejections.[19]

The thematic link *Ladies' World* stories make between earning money at home, authorship, and purchasing power—between writing and shopping—continues into a strategy the magazine used to attract advertisers: it offered cash prizes in a contest inviting readers to write stories from the advertisements, as described in chapter 2. Writing in the contest both let readers earn cash to buy advertised products and brought them into more intimate contact with the advertising.

Ladies' World stories repeatedly proclaimed that earning money gave women a necessary measure of power and control over their own workplace, the house. Women who earned money were entitled to spend it on services and the type of goods advertised in the magazine that would ease or substitute for their work. Moreover, it was clear that earning the money was probably the only way they could obtain such goods. Taking on the familiar formula in which earning money allowed one to buy one's way out of housework and procure new goods and comforts, the Diamond Dyes booklet short-circuited the process and suggested that women should simply increase their household labor—sew more, lug more kettles of boiling water, knit more—to bring more modern and varied commodities into the home. As an individual advertiser selling packets of dye already within the means of its readers, this company's interests were different from those of advertisers as a whole, and different from those of a magazine like *Ladies' World*, which served as the advertisers' representative. Diamond Dyes's adaptation of the formula collapsed it and left out not only the woman's earning of money but also her spending it on advertised goods. The absence of both earning and spending in this exceptional instance highlights their importance to advertisers as a whole in addressing cash-poor and rural women. In general, in order to escape drudgery in the home, such women had to take up some better-paid form of productive labor; writing was presented as one type of such labor.

Writing vs. 𝕸arriage: The 𝕸iddle-Class 𝕸agazine

An attack by a *Ladies' Home Journal* columnist went after precisely the strategy proposed by the *Ladies' World* short stories.

> Lately there has been a great deal of fretful, impatient, womanly writing, about the degrading, depressing influence of household work; and it has been urged that it is better for wives and mothers to write or sew, or do any kind of mental work, in order to make money to relieve themselves of the duties of cooking and nursing. Women who have this idea ought never to have become wives, and they ought never, never, never to have become mothers. For if there is any loftier work than making homes lovely, and sweet, and restful, or any holier work than nursing and training her own little children, no woman will find it in writing, or sewing, or preaching, or lecturing, or in any craft of hand or head known to mortals.[20]

Here, all tasks of housekeeping are fused; cooking and child-rearing are so inextricably linked that the article's logic demands that if the reader wants to hire someone to haul and throw out the bathwater, she must also give up her baby. While some *Journal* readers would have had servants afforded by their husbands' incomes, and would have bought products to substitute for other forms of housework, the real issue was the proposed trade-off of women's mental work for household duties that included the emotional caretaking of men as well as children. "Foolish women," said Barr, are those who "will not stoop to conquer" by coaxing and smiling, but instead "would as soon pet and stroke a Remington typewriter as a stubborn, refractory husband or lover."

Unlike poorer women, the middle-class urban married woman of the 1890s had money to spend on shopping. As she spent less time on the most familiar form of direct production of goods, the time thus made available was the subject of debate. New educational opportunities opened up, and the idea that women might obtain individual satisfaction from work outside the home gained ground, though these opportunities were more often approved for single women. But forces such as the burgeoning home economics movement pushed in another direction: they attached more importance to the household tasks that remained, added more tasks to these, and insisted on a new cult of the home.

Middle-class women's work in the early post–Civil War era had been increasingly defined in terms of the spiritual and emotional influence women should exert over the household, along with the psychological support they were to provide. While this orientation retained its religious overtones, other forms of influence were annexed to the same model by the 1890s. A woman's guiding presence was crucial to her husband and children. Even her neat, untroubled appearance at the table could be a profound influence for good, one *Ladies' Home Journal* columnist asserted. At breakfast, "the men of the household ought to see a woman at her best," pouring out the coffee, "forget[ting] as quickly as possible anything that happened in the kitchen so that its affairs may not furnish conversation for the breakfast table," and looking as "bright and cheery and pretty . . . as a morning glory."[21] Control of her emotions was part of her job. As another *Ladies' Home Journal* columnist explained, "Cheerfulness is as indispensable in the business of being a wife as yeast is in bread."[22] Her responsibilities thus became broader and more nebulous.[23] Ideally, the physical dimension of her tasks should be invisible, barely to be broached in polite conversation at the breakfast table.

Stories about female authorship in middle-class magazines allegorized the surrender of independent work and autonomy at marriage and the taking up of emotional support work. Marriage was often specifically linked with consumption within this shifting cult of the home. Unlike the *Ladies' World* stories, where paid work continues after marriage and where unmarried independence is esteemed, in these stories, authorship is displaced by marriage. In the middle-class magazine stories, authorship offers women the possibility of finding satisfaction and fulfillment as productive workers before marriage. The plots, however, foreclose the continuation of such work and suggest that having and spending money are ultimately superior pleasures.

These stories helped frame the purchasing work of middle-class women less as a job than as a source of pleasure, just as department store palaces of consumption had made "shopping" a form of entertainment. Rather more schematically, from the point of view of the advertising-dependent, middle-class magazines, the full-time housekeeper was a better consumer. Assuming an intact middle-class family with a bread-winning husband, earning money would have distracted from the wife's chief occupation of purchasing goods for the household. To shop at genteel middle-class levels, a middle-class husband's income was a more plausible resource than what most women could bring in through their own work. So it is perhaps not surprising that stories in these magazines generally disapproved of women's earning attempts or subverted them through their plots.

Women writers appeared in numerous courtship stories in the middle-class magazines, but in these stories marriage makes their writing redundant, if only because marriage to the right man would supply a woman's material needs, including her need for goods advertised in the magazine. This trade-off is explicit in Adelaide Rouse's "The Story of a Story" (*Munsey's*, 1896), discussed in chapter 4. Here a woman wants to sell her story to earn money to buy a bicycle; her story is rejected, and her ambition to write is traded in for her engagement to the assistant editor who rejected her story—and for a tandem bicycle to ride with him. So the woman's writing is a poor strategy for buying one of the bicycles advertised in the magazine.[24] Writing is displaced by marriage, and marriage brings in the goods, while the independence and mobility that were part of her original desire are dropped.

The young woman of Rouse's story writes badly, but even when an unmarried woman is a successful writer, her writing leads to marriage. In Marguerite Tracy's "The Unhonored Profession" (*Munsey's*, 1901), writing and marriage are virtually interchangeable. The nameless narrator is a writer courted by Wolfe Hamilton, a doctor with whom she has a comradely friendship but has refused to marry. Wolfe wonders whether he might interest another woman, Leila, in story writing to take her mind off her presumed love for him, since, Wolfe explains to the narrator, "you always tell me, when I ask you to marry me, that you would love me if you weren't too much interested in your writing to think of loving anybody." Leila, who does not in fact love Wolfe, advises the narrator to marry him: "Anybody can write stories—at least there are always plenty of 'em written. You never saw a magazine published empty because they didn't have any stories to put in. But everyone can't look after that ridiculous Wolfe Hamilton, and with his money and position that's worth an intelligent woman's while."[25] Story writing here is something anyone can do; not because it is as accessible as raising berries, but because too many people are doing it already. Marriage to a successful man like Wolfe Hamilton, on the other hand, with its duties of emotional caretaking, takes exceptional talent and will be well remunerated. Marriage and writing fill the same role to such an extent that they are mutually exclusive.

To the extent that marriage and writing were set in opposition in *Ladies' World*, as in "Aunt Crawford's Wise Will," writing was clearly the better alternative. In *Ladies' World* stories, married women could continue to write, and writing was one of a variety of means by which they could earn money. One reason for the contrast between these two types of stories is the expected duties of a farm wife, which included providing for the physical needs of the household, whether by producing goods herself or by earning the money to buy them. The work of the middle-class urban wife, however, was becoming harder to neatly pin down. Central to it was the ill-defined, never-ending labor of providing emotional succor. The narrator of Tracy's story must choose between continuing to make her psychic investment in her fictional characters and devoting her emotional and psychological energies to "look after" Wolfe; she chooses between creating her own characters and building Wolfe's character.

In these stories, writing can also lead directly to marriage, which then displaces it. The fact that a woman once *did* write, once her writing is safely displaced, becomes one of her attractions: it gives her better knowledge of her husband's needs. "The Woman's Edition" by Bessie Chandler (*Ladies' Home Journal*, 1896) relies on this idea: the general incompetence of women writers is a foil for the competence of the main female character; her competence pales beside that of an able and competent man.

In Chandler's story, a group of women inexperienced in publishing put out their town's newspaper for a one-day women's edition benefiting a temperance cause. Because she has no husband and children to tend, Grace Waters, a young college graduate, is chosen editor-in-chief—against her wishes, since she wants to avoid Mr. Terance, the paper's regular editor, who is in love with her. As an educated woman tied to the standards of the "regular" edition and the world of commerce, Grace becomes a touchstone of reasonableness and a conduit for making fun of the bad writing and naïveté of the other women on the paper. She is frustrated by the large quantities of bad poetry submitted by women and the unbusinesslike way in which they prepare it—one woman, for example, arrives with a manuscript tied in different colored ribbons, so that the printers can follow the chromatic order of the colors: violet, indigo, blue, and so on. Grace nonetheless finds pathos in it: "When I think of all these little springs of poetry that these good, hard-working domestic women have been concealing all these years, I could just weep. I'm going to write something about the 'Submerged Sentiment' of middle-aged women."[26]

Other women working on the special edition don't understand at first that, although they are donating their own time, bringing out the paper will entail expenses for paper and press work and that they will need advertising. But as they learn the pragmatic ways of the business world, the women come to accept even liquor ads to earn more money for their cause.[27]

When the paper is finally assembled, Grace breaks down after discovering that the first page is full of drastic mistakes. Mr. Terance heroically stops the presses and fixes everything while she cries. As all is remedied, and as newsgirls gaily sell the paper in the background, she agrees to marry him, finding her match in the real editor. The woman's edition is clearly the inferior version; it appears for only a single day and is produced for charity rather than pay. Work on it, however, leads to marriage, once Grace learns to appreciate Mr. Terance. She now understands enough about the professional life of her future husband to follow his talk and to understand his interests as she supplies emotional support. While Mr. Terance's competence rescues Grace, the kind of support a good wife was expected to provide is different. Chandler demonstrates the dangerous consequences of failing to supply it in her story "A Woman Who Failed." In it, a woman not only lacks faith in her husband's professional abilities, but "could not master and control herself enough to be always a pleasant person to live with"; she thereby blights his emotional and professional life.[28]

While women earning money through writing and other means helped the entire family live more comfortably in the *Ladies' World* stories, and even strengthened the husband's and wife's relationship against challenges by older relations, in the middle-class magazines, writing by married and even engaged women could be downright threatening to the stability of the middle-class household and to the husband's authority and peace of mind. In *Munsey*'s story "Mrs. Medlicott" by E. M. Halliday, an apparent feminist who plans to "continue to edit *The Woman's Friend*" after marriage avoids domestic tasks while overspending her husband's money. By ordering in food sent from the local restaurant instead of cooking for her husband's friends, she forces him to economize—once he gets the restaurant's quarterly bill—by moving to the cheaper neighborhood she prefers: it's closer to her colleagues on *The Woman's Friend*. She thereby takes over the household reins.[29]

The women in the *Ladies' World* stories had only appeared to overspend their husbands' money; the revealed facts of their earnings demonstrated that they had achieved a new, higher economy that allowed them to spend more and raise the entire family's standard of living. They might perform their hidden work to help the husband, as in a *Ladies' World* story in which mother and daughter secretly take in laundry to clear the husband's debts and restore his place in the community.[30] Mrs. Medlicott's work and spending, on the other hand, accentuate her antagonistic relationship with her husband. Working and spending are in conflict, and women's work leads to a wild, destabilizing and unsustainable spending.

Similarly, a story in Frank Munsey's middle-class women's magazine, *The Puritan*, suggests that a woman's writing can make her too independent, and it threatens her mate's security in a sexual and emotional realm. In Matthew White Jr.'s "In the Shadow of Success," Evelyn, as a private project to surprise her fiance Hugh, writes a novel telling their love story, but then sells it for money to take her mother abroad for her health. When it becomes a bestseller, Hugh resents Evelyn's fame and new independence because she no longer needs his protection. Furthermore, the link between their love and her writing taints their entire relationship for him: as Evelyn strokes his brow, " 'She is practicing her art on me,' was the terrible thought that came to him. 'What I do under these circumstances, some hero will be made to do under somewhat similar conditions.' "[31]

Evelyn responds to his resentment by breaking their engagement, saying it is better to recognize their mistake. And she has her work: she has begun a new book. Hugh tells himself that she was brave to have broken contact: "But it is not the brave woman I love. It is the dependent, trustful creature Evelyn was a year ago. It seems exactly as if she were dead." The critics soon agree that Evelyn is not the same creature. Not only is the new work "wordy, pointless, morbid" but also the author "is a bird of one song. This she has sung, and happy for her had she remained forever afterwards mute." Hugh seeks out every one of the scalding reviews, set forth at length in the story. He can now pity Evelyn and see her potential to again be a dependent, trustful creature whom he can console. When an accidental meeting leads him to propose, Evelyn tells him, "I have loved you all the time, but I know that miserable book had changed you, so—so when you gave me the chance I determined to play a part. And I played it well, so well that it nearly broke my heart."[32]

Practicing this kind of art, pretending not to love Hugh in order to bring him back, and perhaps even pretending to write badly, doesn't bother Hugh the way Evelyn's use of their relationship in her writing had. In this scenario, writing threatens to ruin a woman's chances for personal happiness because it breaches a boundary between the precincts of the home and marketplace. The success of Evelyn's writing is tied to its original position as a private gift for her own and Hugh's personal pleasure. When Evelyn takes what was meant for home use and sells it, the results are not as benign as those of the berry seller in the *Ladies' World* story who expands the distribution of her produce from table to market. Evelyn instead has transgressed, and raises the possibility that a woman might be using her and her suitor's love life as raw material for fiction, retailing intimate moments in stories. If her writing is so closely tied to these intimate moments, then she is engaged in a kind of prostitution: providing to the public at large the brow-soothing and emotional succor that as part of her wifely job description should belong exclusively to one man. Women's work, because it brings the public world of work into the tenuously private sphere of domestic work—specifically domestic emotional work—thus disrupts the middle-class family and its privacy.

Support and Discouragement for Women's Writing

So far I have discussed women's writing as a topic within stories. But although periodicals addressed to middle-class women and those addressed to poorer women both printed work by women, they made very different publishing opportunities available. In addition to publishing stories that cast a favorable light on female authorship, inexpensive periodicals for poorer women such as *Ladies' World*, *Comfort*, *The Household*, and *Housekeeper's Weekly* consistently published letters from readers that consituted a lively portion of the copy. Readers even corresponded with one another in print in an arrangement resembling *The Boston Globe's* still-running "Chatters" column (a feature begun in the 1880s). The distinction between letters and articles disappeared as editors of such periodicals asked "aspiring authors" to send in material. For example, Henry Ferris, editor of the five-cent *Housekeeper's Weekly: Woman's Own Paper*, which regularly announced suffrage and Women's Christian Temperance Union meetings, explained in 1890 that the paper

> is mostly written by its readers. . . . I assume that my readers are intelligent women, competent to distinguish themselves between good and bad literature; and they shall have the best that I can get for them. . . .
>
> I wish to have my readers know that I *depend* mainly upon their contributions. . . .
>
> A lady said to me the other day, "I shouldn't think you would *dare* to give your readers the idea that they might *write* for you. Why, *every* one will want to write! Aren't you perfectly *flooded* with contributions from aspiring authors?"
>
> Of course I am, and that is just what I want to be. I *want* every reader to be a contributor.[33]

Contributors were paid. Though *Housekeeper's Weekly* printed few short stories, its articles often included exemplary anecdotes and stories written using the conventions of fiction—pieces that might be thought of as stepping-stones to story writing. The paucity of information available on most of the story writers in *Ladies' World* and *Comfort* (which did publish much fiction) suggests that they, too, were open to publishing unknown, even amateur writers.

The situation at *Ladies' Home Journal* and other middle-class magazines was very different. As "The Women's Edition" suggests, amateur and professional writing were not permeable categories. While Chandler's stereotyped amateur women writers turn out ridiculous poetry, she asserts that the stereotype doesn't apply to her heroine or, presumably, to herself. Chandler's amateur is not a speaking subject, someone with her own subjectivity who can describe her own situation or experience. Nor is she someone who could potentially become such a professional writer: the amateurs in the story instead seem frozen in positions of permanent naïveté and incompetence, forever available as topics for the essay Grace might write or the short story Chandler has written.

The sharp division between amateur and professional writer in the middle-class magazines was consistent with the new segment of the middle class they addressed, with its emphasis on professionalization. While working one's way up the ranks of the working world by amassing knowledge and skills was still a popular explanation of how to get ahead, managers were increasingly likely to come from a white-collar track, with management understood to be a separate, professional set of skills that could not be acquired from the shop floor. Wives of managerial men learned that their work life was as rigidly tracked and that one could not rise from amateur to professional writing by developing skills along the way.

A process Janice Radway has written about, by which present-day romance readers become writers, may cast some light here. These writers not only enjoy what they've read but also identify with the community of romance readers and want to give something back to it.[34] Similarly, turn-of-the-century magazines addressed to rural and poorer women invited readers to see becoming a writer as an uncomplicated transition as they invited readers to contribute their voices to the magazine and become part of a community of writers as well as readers. They could contribute to the community they felt a part of as writers. But while these periodicals allowed entry into writing for an audience by small steps, the middle-class magazines in effect made room only for those who were already professionals. They thus mystified the process of acquiring expertise as a writer and made it less possible for their readers to obtain that expertise as members of the community of readers of that magazine. Writers in the middle-class magazines appear to arrive as outside experts, already trained. Readers of these periodicals were invited to contribute chiefly as consumers.

The *Ladies' Home Journal* of this period has already been the object of considerable scholarly attention, directed at, among other things, the shifts in Protestantism visible in editor Edward Bok's promotion of Henry Ward Beecher's work, the impact on American vernacular architecture of its widely distributed house plans, and its part in Progressive-era regulation of patent medicines through investigation of their contents and support for the Pure Food and Drug Law. Bok's promotion of his own story via his several autobiographies and his biography of his father-in-law, *Journal* publisher Cyrus Curtis, coupled with his long-standing, publicity-generating connections within the daily press, helped make the activities of the *Journal* more visible than those of similar magazines. But in the development of mass-market magazines and the construction of the reader as consumer, the *Journal* occupied a special place.

Under Bok's editorship (1889–1919) the *Ladies' Home Journal* was notably aggressive among the middle-class women's magazines in seeking to serve advertisers: it not only mixed advertising and reading matter but also offered to place ads next to related copy.[35] In well-publicized moves, it excluded advertisers of dubious goods such as patent medicine and financial schemes and warranted the products sold in its pages, thus suggesting to its readers that choices made from among the products advertised within them would ensure safety for the family. Historians have argued that while the muckraking pursued by such middle-class magazines as *McClure's* exposed abuses and made consumers more suspicious of business, it ultimately promoted regulation that allayed fears and strengthened consumers' confidence in business.[36] Similarly, Bok's investigation of patent medicines and his crusade against them provided readers with evidence that the magazine would police its advertising precincts carefully. Although Bok framed his and Curtis's policy against such advertising as a brave one, advertisers agreed that it enhanced the value of the magazine's ad space. "When deceitful advertisements have been finally banished from newspaper and magazine pages, legitimate announcements will have a better chance and will pay larger dividends," one advertising trade journal noted in 1899.[37]

Middle-class women's magazines like the Ladies' Home Journal, The Delineator, and Good Housekeeping took a special and often more self-conscious role in the construction of the woman reader as consumer. Editor Gloria Steinem, in an article introducing the first issue of the newly advertising-free Ms. magazine in 1990, revealed much about the scope of editorial control that advertisers have been allowed to exercise overtly in women's magazines in more recent years. Steinem noted that it has become routine for women's magazines to "supply what the ad world euphemistically describes as 'supportive editorial atmosphere' or 'complementary copy,'" with advertisers' control over the editorial content of women's magazines institutionalized in "'insertion orders' [contracts ordering advertising] or dictated to ad salespeople as official policy."[38] One result is that women's magazines are pressed to generate articles on food, fashion, and beauty subjects.

Insertion orders control editorial content by designating a wide variety of editorial matter a company's ads should *not* be placed next to, as well as, for example, anything "negative in content" (Maidenform) or controversial (S. C. Johnson & Son), or "hard news or anti/love-romance themed editorial" (De Beers diamond company). Proctor & Gamble's demands were particularly extreme, requiring that its products not be advertised in any *issue* that included any material on "gun control, abortion, the occult, cults, or the disparagement of religion."[39] Topics—and space—remaining after the advertisers' demands are fulfilled were obviously much restricted; what was sure of inclusion was inexorably oriented toward consumption.

The *Journal*'s orientation toward service for the advertiser succeeded. It increased ad revenues from $250,000 in 1892 to $500,000 in 1896, and to over $1 million soon after 1900.[40] Its success among the middle-class women's magazines made it a model for others of that type. But it had much in common with the large-circulation mail-order magazines as well: it was cheap, it mixed ads with reading matter, it was roughly tabloid in size, and it attempted to address the interests of the entire family. Under Bok's editorship the *Journal* effectively differentiated itself from these mail-order magazines and came to be grouped with the middle-class magazines. Bok's press-agentry helped in this redefinition, and his campaign against patent medicines might be classed in this category. It was Bok's willingness to pay high prices for stories and serials by such authors as Kipling and Howells, and to publicize their prices, that made the *Journal* unmistakably a player in the field of middle-class magazines.[41] As Bok configured the magazine, its prestige and his own depended on publishing well-known professional writers.

Ladies' Home Journal columnists fulminated especially against the notion that amateur women might make money from their writing or editorial projects. Bok, for example, made fun of the idea that women might earn money by compiling literary extracts or bits from scrapbooks or by selling occasional verse. And yet such projects could be a literary apprenticeship, in which a contributor would become accustomed to the demands of publication—learning, for example, not to send in poetry keyed by different colored ribbons. Such a contributor would also learn to persevere through rejections—something the *Ladies' World* story "Aunt Crawford's Wise Will" showed was necessary.

While Bok's mockery of the attempts of amateur women writers could be understood as an endorsement of a higher level of professionalism for all writers, it mystified the process, denying links between tentative and more ambitious writing projects and eliminating the layer of apprenticeship available in magazines addressed to poorer women. Ironically, Bok's autobiography shows that he developed his own writing from similar small beginnings.

On the one hand, Bok simply seemed to be enforcing the idea that men should earn while married women should take charge of consumption. But his assault on women's writing went further than attacks on attempts to write for pay. Having read what he said were thousands of letters from readers, Bok launched an 1893 attack on women's letter writing. He complained of receiving "indiscreet" letters from women, but at the same time asserted that women's letters were both more "natural" than men's and, therefore, more "delightful," since "a man always writes a letter by a certain formula; a woman ignores all formulas and writes as she feels." In a period that, as we will see, increasingly came to value terseness and succinctness in writing as more businesslike, this "delightful" writing, with its "inevitable postscript which so often says more than her whole letter," was markedly marginal.[42] Women's supposed letter-writing practices, though Bok claimed to cherish them, became evidence for the incompetence of women to write outside the home sphere.

Why was Bok reading so many women's letters? Much of the popularity of the *Journal* was credited to the columns devoted to answering readers' letters, called readers' departments. Thousands of readers wrote queries to the *Journal*, and magazine workers answered them. While this service seems to have inspired readers' loyalty, it strikingly avoided creating the kind of community of readers that prevailed in magazines such as *The Household*, *Comfort*, and *Housekeeper's Weekly*. The readers' departments filled the *Journal*'s pages with unpaid reader-generated material. Perhaps a quarter of some issues' editorial matter was made up of answers to readers' letters—in columns on child-rearing, etiquette, religious activities, and tips on appearance. The fact that the letters themselves rarely appeared reinforced the sense that women's voices should not be heard in public and that the editors must protect women from speaking indiscreetly. "If sometimes the heart rules the pen, instead of the mind, it is more than ever incumbent upon the recipient to remember the confidence imposed, and respect it," Bok said of women's letters.[43] The result was that women's words were mediated through the magazine's control: readers received only the oracular advice of the columns' editors, rather than learning directly about other ordinary women's lives—perhaps through messy, indiscreet letters. Moreover, unlike those who wrote to advice columns in the magazines for rural and poorer women, the letter writer's contributions to the *Journal* became invisible—and naturally, invisible writing was not worth paying for. The writer became simply a consumer of advice. The real occupation of the middle-class *Journal* reader was shopping, not writing, not earning money.

Bok's account of his campaign against patent medicines attacked Lydia Pinkham's Vegetable Compound with particular gusto, proudly recounting his publication of a picture of her grave, even as the ads continued to advise readers to write to Mrs. Pinkham with their problems. The vitriol he directed toward Pinkham's business makes more sense in relation to the advice-seeking letters the Pinkham company solicited than its relatively harmless compound. One function of the *Journal*'s readers' departments was to foster a sense of dependence by readers. The magazine was an ear as well as an oracular voice which shaped the kinds of concerns readers would address in the letters, and which shaped their relationship to the magazine: it was a trusted listener as well as advice giver. Pinkham's letter-answering service was in direct competition with the *Journal* as a source of information and sympathy.

The *Journal*'s strategy was both to discourage women's action in the public and entrepreneurial realms and to encourage lives that placed consumption at their center. Shopping, within this structure, was an acceptably *social* rather than transgressive *public* activity.[44] Writing, like other forms of public action, was antithetical to caring for a home, Bok explained in an 1894 column advocating a "practical" education for women in "the great art of housekeeping":

> We may admire the public singer on the concert platform; we are charmed by the actress on the stage; we are impressed by the woman who writes well or talks brilliantly. But, after all, the woman who *holds* us, who not only commands but retains our respect, is the woman who is truest in her own sphere, reigning over her kingdom of home and children with a grace and sweetness, compared to which a public life is the hollowest of mockeries.[45]

Bok thus flattered the women he assumed made up most of his readership: women at home, living on a husband's income, without a higher education, and who were comforted by assertions of the psychic importance of their work. While ostensibly arguing for girls' education, his article more carefully worked to reassure readers that their current way of life was best. So while the *Journal* appears to present-day readers to seethe with contempt for women, readers in Bok's era may have found support in it.

One reader, Magnolia LeGuin, a white farm woman living something of a hardscrabble existence in Georgia—her family had enough to eat but little to spare—writing in her diary in 1902 when she had three small children, called the *Journal* her "favorite magazine . . . how much pleasure a subscription is for me I can never express." By 1910, expecting her eighth child, she no longer subscribed, but looked forward to a neighbor's loan of a year's worth of copies "if I can find to read them. . . . Words fail me when I try to tell how I enjoy those magazines." She regarded the *Journal* as a source of strength for the coming ordeal of childbirth and a point of contact with other women's lives in a period when she records having "taken two meals away from home this summer—first meals in five years." LeGuin did not seem to choose the *Journal* over another women's magazine; it was simply one she knew about, and the *Journal*'s campaigns for subscribers may have disseminated it more thoroughly than other magazines. Its reprints of sermons and its pious tone may have also placed it in the register of religious reading that LeGuin already found acceptable. It sometimes served her as a conduct guide, a supplement of sorts to the Bible she read every night. She copied down its precepts on not speaking evil of others, for example, not remarking on the fact that one of them was framed as "one of the first secrets of popularity." In another year, a subscription, at $1.50 a year, was a "rich treat," and took its place with other Christmas presents like a "new pretty cake dish." The magazine seems to have served LeGuin as an ordered, genteel world to retire to in the few moments she had for it.[46]

If the magazine allowed a moderately poor farm woman like LeGuin to see herself in its genteel pages, other, less marginal readers may have found its world even more familiar and accessible.[47] The strategy of idealizing middle-class life, holding forth the situation its assumed readers were in as the ideal one, was developed in many stories in which a woman character was responsible for choosing her economic condition, a condition that very likely resembled that of many readers. In this formula, a rich woman chooses a poor man; she must convince him that she really wants him and is willing to be poor along with him. Identifying with the woman in the story who chooses such a marriage and even has to fight for it may have allowed readers to see themselves as having chosen the economic situation of their own lives, which probably included financial constraints. The woman's choice in such stories, too, perhaps allegorized leaving a middle-class family home and marrying any nonwealthy young man, stepping away from the accumulated comforts of an established middle-class household into that of a couple just starting out.

Integral to such stories was the agreement that the couple would live on his income alone, along with the reinforcement of the belief that it would be shameful to use her income either from inheritance or earnings. This agreement becomes essential to the proper resolution of a romance; in fact, the stories show such agreement as something women ardently desire, and achieve only after great effort.[48] But although the men of these stories are poor, they are typically on the rise in white-collar positions. The emotional work of helping them, and of giving them a reason to succeed, takes precedence over raising cash for immediate shopping.

Other stories further accentuate this element of the face pressed longingly against the windows of domestic life, of the woman who achieves the life of a middle-class housewife only after overcoming some obstacle. Some ordinary element of middle-class household life is celebrated, from the point of view of someone who is excluded from it or prevented from enjoying it for reasons other than poverty. For example, in Juliet Wilbor Tompkins's "His Dutch-Treat Wife" (*Ladies' Home Journal*, 1905), Olive is excluded from the pleasures of shopping and housekeeping, held forth as a wonderful delight. Her theories of independence and the couple's apparent need for money keep Olive at her job in her "dear, exciting office" after marrying Ernest. Olive also takes on complete responsibility for running the new household, believing she can do it in fifteen minutes a day. (There is no hint that Ernest might do any of the work.) But housework seduc-

tively absorbs more of her attention. She is preoccupied at the office: "Olive, hith-
erto whole-souled in her devotion to her work, was annoyed nowadays to find
her mind slipping away in the late afternoon to trivial housekeeping details." Her
initiation into the mysteries of shopping leaves her especially distracted: "She
even arrived late occasionally, having lingered over seducing piles of fruit and
vegetables, or some marked-down display of fine table-linen, in which her eyes
seemed to be daily opening to new distinctions." When her friend Florence shows
her a new embroidery technique, she rouses Olive's inborn longing to do fancy-
work; Olive stays up all night with it: "'I can't help it—it's an obsession,' she apol-
ogized. 'I think of it all day; I fairly run home to get to it.'" Her discontent emerges
in what the story asserts is a primal, instinctive desire to quit her office work: "the
encroachment was so often from within rather than from without." "'Why
shouldn't I have what Florence—what every other wife has?' might have been the
expression of it, if she had allowed it words. She hated herself for the pettiness of
it—not dreaming that she was struggling against the mighty force of tradition."
When her husband returns her financial contributions to the household as a
birthday present, expressing concern about her fatigue, she gratefully leaps at the
chance to back down from her earlier statements, give up her job, stay home, and
begin a family: "I am so glad to stop work. I—I don't want to be an individual
now," she declares. What "every other wife has" is said to be both what Olive naturally and instinctively desires, and a valuable prize, achieved after deprivation
and difficulty. The story's resolution leaves her gratefully attaining the position of
full-time housewife in which most of the *Ladies' Home Journal*'s readers were presumably already ensconced.[49]

Bok's antipathy toward women's enterprise stands in sharp contrast to his
own youthful entrepreneurship, as extolled in his autobiography and in his celebration of the rise of his father-in-law, Cyrus Curtis, to the ownership of the
Ladies' Home Journal and the *Saturday Evening Post*. He praises himself and Curtis for seizing on all means of converting scraps of writing, odds and end of
tasks, into cash. But such enterprising vigor is not praiseworthy in women, who
are relegated to a suitably complementary position in Bok's autobiography: they
provide occasions for Bok's entrepreneurial genius to burst forth. It is important that women be economically dependent because it is finally from their
dependence, and from their adherence to their work as consumers, that the
advertisers Bok represented could benefit.

As we have seen, the *Ladies' Home Journal* promoted the idea of a very separate, professional caste of writers. And yet perhaps one reason for its success
was that the *Journal* as a whole was not so simple or single-voiced. The same
magazine that routinely disparaged women's writing also offered advice on how
to mail a manuscript. And although in its stories women's attempts to earn
money ended badly, the magazine for a time ran a column to which women
contributed ideas for earning "pin money."

Ads were one source of this multiple-voiced quality of the *Journal*.[50] While
editorials and stories ridiculed women's suffrage, the "new woman," or the idea
of women serving on juries, ads might appropriate the excitement of new political possibilities by advising, "Don't sweep the old way! The New Woman

Figure 5-1 *Ladies' Home Journal*'s ad for a carpet sweeper caricatures the backward broom user and aligns the product with exciting new political possibilities, even as it reduces suffragette to sweeperette. (*Ladies' Home Journal*, January 1896, p.41)

Sweeps Hard and Soft Carpets, Bare Floors, with a *Sweeperette*"[51] (figure 5-1) or show a woman lawyer addressing women jurors, charging them " *'You must decide* that the "S.H.&ʳ M." Bias Velveteen Skirt Bindings *are not guilty* of any of the defects charged against other bindings. The verdict must be: *The Best Made*"[52] (figure 5-2). These feminist and quasi-feminist catch phrases and slogans patronizingly trivialized a serious quest for political power into a choice of trimmings and appliances and suggested, along with anti-suffrage propaganda, that a woman's effect on the home sphere was so powerful that it exceeded and made unnecessary any power she sought in the larger world. And yet a reader with feminist sympathies might have found the ads appealing. Certainly her eye might be caught by the slogan, and here, the women associated with these slogans were, for once, not the ones caricatured. Women appeared as jurors, in a powerful position denied them in fact, and are shown as attractive and well dressed. The ads might, then, have appeared to offer the reader a more progressive arena than the rest of the magazine. The predominant conservatism of the rest of the magazine reassured, however, that middle-class order would not be thereby challenged or disrupted.

While a John Kendricks Bangs story in the same issue mocked the qualification of women to be jurors by showing one abandoning her case to rush home for a household emergency, the S.H.&M. ad hinted that housework and jury service were compatible and that evaluating and choosing goods might even be worthy of respect.[53] The kind of women's work Bok's editorials were likely to praise was the the nebulous business of "reigning over her . . . home and children with grace and sweetness";[54] they less often focused on more con-

Figure 5-2   A woman lawyer and jurors
attract attention in this *Ladies' Home Journal*
ad for skirt bindings. (*Ladies' Home Journal*,
December 1895, p. 31) (Courtesy of the
General Research Division, The New York
Public Library, Astor, Lenox and Tilden
Foundations.)

crete details like housecleaning and shopping. Maintaining her untroubled,
morale-building appearance at the family breakfast table depended on excluding from conversation the affairs of the kitchen; as we have seen, decorum
required that she even "forget as quickly as possible anything that happened in
the kitchen."[55] But even middle-class women spent much time in the kitchen,
on cleaning, sewing, and choosing skirt bindings, and the ads were one place
they might see this work reflected.

On the other hand, the theme of choice-making visible in the jury ad, so
much a part of the newly formulated work of shopping, was central to many
stories. Shopping itself involved decisions with moral weight. Home decorating choices, for example, had moral force, since, as one interior decorator
aligned with the House Beautiful movement put it in 1893, "A perfectly fur-

nished house . . . not only expresses but *makes* character."[56] While such a claim came packaged with a set of Arts and Crafts Movement decorating suggestions, even those with looser notions of what the beauty that would shape the character of its beholders would look like accepted the broader claim. Making a beautiful and comfortable home was the first step in keeping husband and children at home and safe from vicious outside influences.[57] It could even rescue men in danger of going wrong, as in one 1890s story of a woman who took up the mission of running a gentlemen's boarding house, making it a "home" and steering the boarders from bad company by providing easy chairs, a piano, and a pretty lamp.[58] Purchases chosen for the home, then, had great importance for the moral well-being of the household.

Courtship and Shopping

In enforcing the consumer role and either disparaging or making other possibilities for women invisible, the *Ladies' Home Journal* supported women in some of the new tasks of steering through life as a genteel consumer. Stories in middle-class magazines valorized a life of economic dependence rather than independent earning for women. Courtship stories were a staple of these magazines, as they had been of much fiction before. But a theme emerged in some of these courtship stories of flattering women's choice-making in much the same terms as ads did. Women characters chose between suitors: they shopped wisely.

While stories rarely featured scenes of actual shopping, shopping filtered through the magazines in various forms.[59] In refashioned presentations of thrift, magazine articles reinterpreted taking time to shop carefully and learn about goods as wise judgment. Choosing foolishly appeared though stories and columns that ridiculed women who had the bad taste to wear flashy ribbons and hats: their cheap finery was not simply a class marker but a sign of incompetent shopping. Reading carefully through the ads was one form wise judging and shopping might take. And, as we have seen, magazines encouraged careful ad reading through contests and other incentives. By extension, a rational shopper might consult the good advice of a magazine she perceived as a trusted friend—perhaps a magazine like the *Journal*, which had demonstrated that it cared enough about her welfare to exclude patent medicine ads.

Magazine references, in other words, focused on the consumer's social and practical uses of goods as the basis of choice making. By contrast, an available model of shopping which did not appear in magazines was one that asked shoppers to base their decisions on the working conditions of the goods' makers and salespeople. The Consumers' League, begun in 1891, suggested that consumers direct their attention to the conditions of production and thereby exercise their power of economic choice-making to force changes in these realms. As we will see, this approach was frowned on in courtship stories.

In the newly advertising-dependent middle-class magazine, a form of the courtship plot in which a woman chose between two or more suitors mirrored the situation of the middle-class married woman reader, whose work, to an increasing extent, consisted of choosing between alternative products by brand name as she shopped for her family. Within the discourse of the magazine as a whole, the evident disparity between choosing a husband and choosing a cereal was collapsed from both ends: the advertisements warned that choosing the wrong product could bring embarrassment, impurity, disease, and death into the family (see figures 5-3 and 5-4), while the stories showed the choice of a husband being made over and over, as often as the formula appeared in the magazine.[60]

Judith Williamson has noted that late-twentieth-century advertising "sells us 'choices': or, to be more precise, sells us the idea that we are 'free' to 'choose' between things. To nourish this 'freedom,' advertising must, like other key ideological forms, cover its own tracks and assert that these choices are the result of personal taste."[61] In these magazine courtship stories, women's ability and freedom to make their own choice of mate were celebrated; economic information and family influence appeared to step aside, allowing a woman's heart and taste

(My mamma used Wool Soap.)   (I wish mine had.)

Woolens will not shrink if

Wool Soap

is used in the laundry

Wool Soap is delicate and refreshing for bath purposes. The best cleanser. *Buy a bar at your dealers.* Two sizes: toilet and laundry.

RAWORTH, SCHODDE & CO., Makers, CHICAGO
3 Chatham St., Boston.
63 Leonard St., New York.   927 Chestnut St., St. Louis.

Figure 5-3   Unwise shopping choices have consequences for the family, just as an unwise matrimonial choice might leave a woman and her children inadequately clothed and fed. (*Ladies' Home Journal*, February 1896, p. 35.) (Courtesy of the General Research Division, The New York Public Library, Astor, Lenox and Tilden Foundations.)

Figure 5-4   The health of the family is at stake in the grocery store as well as in the kitchen, according to this ad for shortening. (*Ladies' Home Journal*, November 1897, p. 39.)

to guide her to a husband. Hearts and taste however, are the better-assimilated versions of economic considerations, class, or family advice; they have been so well absorbed that they are no longer accessible to articulation, and appear as "natural" or "genuine" feeling. As we saw in chapter 2, advertisers themselves tied shopping to marriage: a woman was newly minted as a consumer at marriage, and the advertising she had absorbed in childhood would have impressed itself so deeply upon her that she would not be able to examine it logically or question it; it would override family buying habits. Women were praised in both stories and ads for their apparent ability to choose both commodities and husbands seemingly independent of the influence of family and friends. The stories played out several scenarios linking shopping and courtship.

Allegories of Shopping

The analogy between choosing a husband and shopping for goods was presented as a joke in an anonymous verse, "A Substitutor's Obituary," in a 1903 advertising journal. It attacked the retailer's practice of "substituting" non-brand-named merchandise for the advertised product for which the consumer, made aware of distinctions between products by advertising, had presumably asked:

> Here lies the clay of Druggist Brown,
>   The Substitution Fiend,
> From selling imitation goods
>   His livelihood was gleaned;
> Let not his mourning spouse be doomed
>   To life-long widowhood,
> The matrimonial market offers
>   "Something just as good."[62]

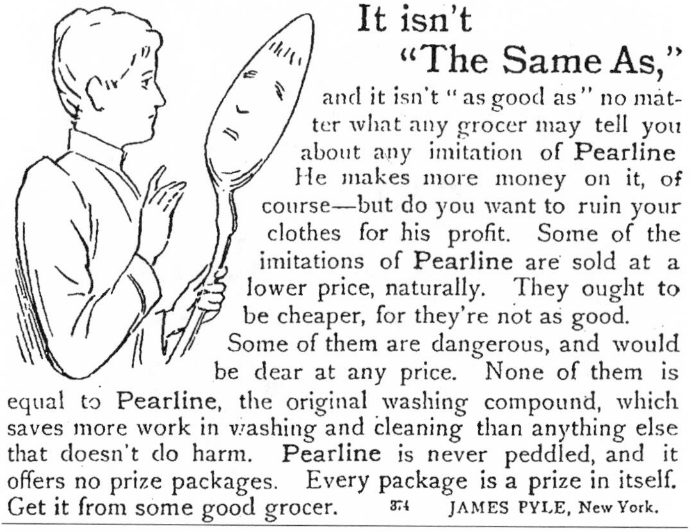

It isn't
"The Same As,"

and it isn't "as good as" no matter what any grocer may tell you
about any imitation of **Pearline**
He makes more money on it, of
course—but do you want to ruin your
clothes for his profit. Some of the
imitations of **Pearline** are sold at a
lower price, naturally. They ought to
be cheaper, for they're not as good.
Some of them are dangerous, and would
be dear at any price. None of them is
equal to **Pearline**, the original washing compound, which
saves more work in washing and cleaning than anything else
that doesn't do harm. **Pearline** is never peddled, and it
offers no prize packages. Every package is a prize in itself.
Get it from some good grocer.   574   JAMES PYLE, New York.

Figure 5-5 Ads warned women against both products that were not
nationally advertised and against the grocers who promoted them,
threatening that such products were "dangerous." (*Munsey's*, April 1893)

Druggist Brown gets his posthumous comeuppance for trying to sell goods
more profitable to him in preference to the less profitable advertised products:
he turns out to be as interchangeable as he claimed the goods are. His widow
holds power to take up with shoddy matrimonial merchandise and besmirch
his name by declaring the new husband "just as good." (The verse turns on the
same anxiety about confusion between relationships to mass-produced products and to people as in Amelie Rives's *The Quick or the Dead?*, discussed in
chapter 3; here it's expressed in jocular fashion.) However, the more typical version of the courtship-as-shopping story echoes such pitches as the Pearline ad
in figure 5-5 and the Pillsbury ad in figure 5-6. It honored the consumer's
power to refuse to buy "something just as good" and to insist instead on the
genuine, advertised item. In this reformulated vision of the marriage market,
women held power; a man's happiness was entirely up to women's wise decision making.

Eighteenth- and nineteenth-century British and American literature
abounds in representations of a marriage market in which women are
exchanged, bartered, or sold. Such representations were implicit critiques of
the situation, as in George Eliot's *Daniel Deronda* (1876) or Edith Wharton's
House of Mirth (1905, serialized in the elite magazine *Scribner's*). The woman on
this traditionally framed marriage market was, as Rachel Bowlby points out of
the prostitute, both seller and commodity.[63] Wharton's Lily Bart, for example,
imagines she can cut a good deal for herself, can trade well, but in fact as a

Losing a Good Customer.

Everything is measured by a standard, whether it be something to eat, to wear or to use. The standard of flour is Pillsbury's Best. No one tries to make any better flour, no one claims that there is any better flour, but some dealers offer for sale other flour which they claim to be just as good as Pillsbury's Best. Why not insist upon having the standard, and avoid the substitute?

Pillsbury's Best Flour

is the best for you and it is the best for those dealers who keep their trade year in and year out by satisfying their customers; but it is not the best flour for those dealers who do not care what they sell, so long as the profit is big. You may be sure that your interests are not thought of when another flour than Pillsbury's Best is recommended.

Pillsbury's Best Flour is for sale by grocers everywhere. Being the best, it is imitated, and consumers are warned not to accept substitutes. "The Best Bread," a book of bread, cake and pastry recipes, sent free.

Pillsbury-Washburn Flour Mills Co., Ltd., Minneapolis, Minn.
Makers of Pillsbury's Oats and Pillsbury's Vitos.

In answering this advertisement it is desirable that you mention MUNSEY'S MAGAZINE.

Figure 5-6 The virtuous woman shopper refuses the blandishments of the grocer and insists on the advertised product. (*Munsey's*, September 1900)

trader she proves to be too easily distracted by other considerations. She finds her value as a commodity diminishing; her speculations are unsuccessful. Lily mistakenly believes that she has the power to choose the situation she wants; the novel presents the frightening specter of her powerlessness as each chance drops away from her, as she reverts to her place as goods, and as men make choices around her.

Critiques of the marriage market in the ten-cent middle-class magazines were usually more simplistic. In "Her Triumph" in *Munsey's*, Thomas Winthrop Hill explicitly called a debutante ball a market:

> She sat like a queen looking down at them all—
> While seven gallants bent before her—
> For she was a debutante at her first ball,
> and the seven were there to adore her. . . .
>
> No wonder she blushed! For a blush she had need;
> 'Twas not pride in her triumph that drew it.
> She sat there a slave to ambition and greed—
> A chattel, *for sale*, and she knew it. [emphasis in original][64]

The debutante's definitive position as a commodity or chattel seems to place her within a larger sales system of which she is not in control, but the shame that attaches to her is personal. She is to blame for being enslaved to ambition and greed. This is not, in other words, an attack on a system of sale, as abolitionist slave market images were a few decades before. The debutante ball setting of this critique, however, within a middle-class magazine, specifically labels the marriage market as an upper-class depravity. The middle-class magazines' characterizations were similar to Thorstein Veblen's analysis of "leisure class" use of women as indicia of their husbands' wealth. Stories in which upper-class women married, or attempted to marry, for money and ended badly were something of a staple of fiction in magazines such as the *Ladies' Home Journal*.[65]

But this grim vision was largely absent from stories set within a middle-class milieu in the middle-class magazines. Instead, the traditional notion of the marriage market was sidestepped and the female character became the one with agency. The roles of seller and of commodity were displaced: the female protagonist took the seemingly powerful role of chooser or shopper, a role that flattered the woman magazine reader, who was also a shopper. The stories suggested that the endlessly repeated work of shopping and choosing was each time of enduring significance and that her role as shopper gave her power.

One 1898 *Munsey's* story framed the situation succinctly: "Two men were in love with her; both had offered themselves, and now the question was, which?"[66] To choose wisely between her two suitors, Miss Littlefield 'must look at it from all sides,' she mused. 'I have always declared that I would never let my fancy run away with me, and I won't—I won't!'" While Clyde and his rich family are altogether correct and of her own set, she loves John, the poorer suitor, but is afraid to decide on the basis of love since "it must be uncomfortable to be always fearing one's fiancé would do the wrong thing," a position John might land her in. She imagines John's mother distressingly fat, in a "calico wrapper

. . . rocking and looking out the window." But meeting John's entirely presentable mother, finding her properly dressed in "a gray gown, with soft, old
lace at the neck and sleeves," ready to greet her in a warm but genteel manner,
settles all doubts, and allows her to choose John on the basis of both heart and
sense. The story suggests, then, that the first impulse to buy is not a bad thing:
when Miss Littlefield gets the kind of detailed information from one side that
advertisers believed women wanted, she learns what John's family's true class is
and proves her fears and hesitations groundless; the rational evaluation "from
all sides" that would have joined her to Clyde is not the best shopping strategy.[67] And yet because her information is coded in terms of taste, and the signs
of realism of the middle-class magazines, the story demonstrates that the heart
or womanly judgment—the thoroughly assimilated, inarticulable version of
knowledge about goods—leads to proper choices.

Though magazines might flatter readers for their wisdom in choosing from
the heart, they also mocked women for bringing criteria from outside the home
sphere into their decision, as the Consumers' League had advocated. In Anne
O'Hagan's 1909 *Good Housekeeping* story, "Rhodora, Advance Agent of the Better Day," Rhodora, a high-minded college graduate, visits her flighty friend
Clothilde. Clothilde is wooed both by worthy Dr. Sparling, who "doctors all the
mill people for nothing," and by substantial mill owner Dwight, whom she
favors. But Rhodora, who "look[s] into industrial conditions" during her visit,
opposes the match and attempts to make Dwight's industrial practices the basis
of Clothilde's choice, pointing out that Dwight "doesn't do *one single thing* for
his mill hands that the law doesn't force him to do."[68]

The standards Rhodora wants her friend to use in choosing a husband parallel those of the National Consumers' League. As the advertisers, ad-dependent
magazines, and Rhodora would agree, shoppers had economic power. Consumers' League members argued that shopping choices could be a force to
change industrial conditions and encouraged shoppers to choose goods based
on the working conditions under which they had been produced rather than
according to price or other criteria. Florence Kelley, a Consumers' League
leader, declared that because consumers drive production, they must therefore
use their power carefully, with a larger good in mind. The League was to "moralize this [purchasing] decision, to gather and make available information
which may enable all to decide in the light of knowledge, and to appeal to the
conscience, so that the decision when made shall be a righteous one."[69]
Rhodora has attempted to moralize Clothilde's decision, gathering for the hus-
band-shopper the sort of information on industrial conditions that the League
provided to department store customers.

But Rhodora's gesture toward power in shopping is held up to ridicule: it
pushes women into the male sphere of industrial decision making. The story
finally demonstrates that the studious bluestocking Rhodora has no grasp of
the real world. Rhodora's disapproval only temporarily deters the marriage, and
when Clothilde finally elopes with Dwight she telegraphs Rhodora that her new
husband has promised to undertake "all those things. You know—the reforms."
When Clothilde's father reads the telegram, he protests: "'Has the little goose

been mixing up in business?' he asked. 'And after poor Dwight kept the mill going full-handed and full time, at the Lord knows what drain on his personal resources, all through the panic! The Puss had better stick to her knitting!'"[70]

Rhodora is finally humiliated, both for having brought the wrong criteria to bear on a shopping choice and for having misjudged the industrial situation. Manufacturers and potential husbands *are* doing what's best for their workers; these are not questions shoppers should pay attention to.

Clothilde exercises her power as a shopper in choosing between suitors, embracing the idea that the shopping choice should evaluate only the goods themselves and the relationship between goods and consumer, not the relationship between goods and producer or that between seller or promoter. Similarly, *Good Housekeeping* promoted its own acquiescent version of consumer power in its Good Housekeeping Institute department, which praised advertisers' products, sometimes including readers' reports of positive experiences with them. The magazine thus presented advertiser-driven information as a form of consumer power. It reconfigured the mail-order magazines' provision of a forum for readers to write in as well by granting readers such space only for the purpose of praising advertised goods.

In the most schematic of the stories reviewed here in which choice-making itself is overtly featured, Lulu Judson's 1896 "A Girl's Way" in *Munsey's*, a young woman chooses between a rich and a poor suitor: " 'I must decide,' she thought. 'I suppose I am just an average woman, but it does seem strange that I do not know which of these two I like the better, or if I really love either of them.'" A stream of family and friends press her to marry rich Mr. Dillard, and she seems to lean toward him. As long as his wealth appears to her both desirable and vague, he remains an acceptable suitor. But when her twelve-year-old brother falls off his old bicycle and complains "It ain't fit for a fellow to ride. I do hope, Ellen, when you marry Mr. Dillard, he'll give me a decent wheel," Mr. Dillard's wealth and the meaning of marrying for money are made concrete, and she rejects him.[71] As a wise shopper, it's not that she thinks money is not an issue. Her poorer suitor, the dynamically named Mr. Wheeler, is a lawyer with drive and ambition, and she assumes he won't remain poor. But in this conversation with her brother she stops being the shopper: her position is too obviously revealed as a commodity, equivalent to, and therefore barterable for, a bicycle, an object whose position as a commodity is underscored for the reader by the advertising for it in the back of the magazine, with the price available for reference as well. In this scheme, she is at best a means for acquiring commodities, a conduit for the wishes of others. She chooses the position in which she will not remain a commodity, but will be the buyer.

Summary

The middle-class magazine featured a cornucopia of products in its ads. While the magazines addressed to rural and poorer women carried more ads for goods to be ordered directly from the manufacturer by mail order, and included numerous ads inviting the reader to act as an agent selling products, suggesting an easy move between buying and entrepreneurial selling, ads in the middle-class magazines more often featured brand-name products that readers could see and buy at stores near them. Ads provided a kind of background for a purchase, and because ads for products like Pears' soap, Pearline, Scourene, and Sapolio ran in every issue, ads gave the products an ongoing life within the two-dimensional neighborhood of the magazine, as well as in the three-dimensional world of the grocery shelf.

Readers of periodicals like the *Ladies' Home Journal* that excluded ads for dubious products "have more confidence in the ads appearing" in these magazines and newspapers, and therefore made better customers.[72] Other women's magazines took up this policy, and, as the advertising manager of *The House-keeper* noted, "to the extent that their advertising columns have become trust-worthy, in the same degree have they become well filled"; moreover, ads placed in "trustworthy" advertising columns like his got larger returns, he argued.[73]

Women chose among these trustworthy suitors admitted to their homes and were flattered as people who make the right decision among both men and products. Next we will see the other side of the courtship story, as admen woo consumers.

6

"Men Who Advertise":
Ad Readers and
Ad Writers

*For ten or twenty-five cents one can buy an attractive monthly trade
review of 150 pages filled with well-written and enticing descriptions of
household articles and novelties duly priced and pictured. The
monotony is relieved by a few pages in the middle of the volume
containing short stories and some half-toned gossip on "American
Beauties and How They Dress," or "Celebrities Kinetoscoped." These
are very nice to turn to when one is tired of looking at things he cannot
buy, but the center of gravity of the periodical has passed from the
literary portion to the advertising.*

editorial, *The Independent*, 1902[1]

The Place of Ads in the Magazine

The inclusion of ads, bulky and clearly economically important to the magazine, excited both approval and annoyance. Were the magazine ads valuable
and useful, as essential to the experience of reading a magazine as they were to
the economics of publishing it? Or did ads impart a negative value to the magazine, making it worth less to the reader than a magazine without advertising?
Since a magazine with more advertising was likely to cost the reader less than
one with little advertising, did that in itself mean it was also less valuable?

Commentators writing in advertising trade journals and in advertising-
dependent magazines were hardly disinterested parties. They did not simply
assert that ads were valuable to readers, but rather attempted to persuade their
readers that reading ads was good for them and that an ad-filled magazine was

more valuable than a magazine without ads. Such essays rationally argued advertising's benefits to readers. But like advertising contests, they addressed their advertisers at least as much as the readers, assuring them that the readers were learning the value of advertisements. Advertising trade journals, sometimes produced by nascent advertising agencies, more directly addressed potential advertisers, assuring them that readers not only read the ads but found them as worthwhile to themselves as the advertiser did, and that everyone joined in celebrating the value of ads.[2] Additionally, as middle-class magazines started to move advertising out of their specially zoned sections in the front and the back, these essays set out to reassure readers that the presence of ads wasn't ruining the neighborhood or reducing the worth or quality of the magazine.

The objections to advertising's prominence in magazines were summed up in a 1903 editorial in *Current Literature*, a literary journal whose own ads were mainly from book publishers rather than brand-name commodities. It held that ads were intrusive and out of place amid high-art literature; they were ruining the neighborhood. As ads move into the body of the magazine, "few things are more annoying to a lover of books than finding a whole page in the middle of an interesting article devoted, let us say, to the advertising of some special shoe, or semi-digested patent-food."[3] The word "book," though standard trade usage for the physical magazine, implied that a magazine should be a literary artifact divorced from commerce. If a magazine ran ads alongside its articles, it was no longer like the more prestigious and permanent book.

Even if some readers do want to learn about the products advertised, their rights should not prevail over those who don't, the editorial suggests. The rights of advertising and literature compete as well: advertising may be an important art in the conduct of business, but "it is not reasonable that the art should be foisted upon the face of the literature of which magazines are such able exponents." Subscribers' rights are disregarded. Not only do ads interrupt reading, but also the subscriber who wants to treat the magazine as a book, to bind issues and keep them, is obligated to keep the "pages which are utterly useless and a positive disfigurement to the volume." Moreover, "the presence of advertisements among legitimate matter of the magazine depreciates its value without increasing its usefulness; imparts to it a decidedly ephemeral character; and lowers its dignity to the level of the daily sheet whose final fate it may be to be wrapped around a pound of nails in some obscure country store."[4]

Advertising was seen to thus impart negative value to the magazine. It not only lowered its price but also reduced its worth and its prestige as well: time-bound rather than a vessel for literature for the ages, issues were no longer worth saving. Ads pulled literature down to the realm of the service-oriented and commercial and reminded readers too strongly of such matters when they could be disengaged from thoughts of production and consumption. In a resoundingly unprescient conclusion, the editorial looked forward to the time when advertisements would be relegated "to the part rightly devoted to it, and so leave us a body of permanent literature which can form an artistic, as well as valuable, record of the nation's progress and literary tastes."[5]

Reassuring the Advertiser

Ad-dependent magazines responded to such attacks and implied attacks by insisting on the value of ads. *Collier's* advertising manager E. C. Patterson's 1909 essay "The Cost of Advertising" seems intended to reassure advertisers while it pleads with readers to pay attention to advertising. *Collier's* advertisers may have especially needed reassurance, since the magazine had in the same year crusaded against patent medicines, perhaps stirring advertisers' apprehension that readers would become hostile to advertising claims. Patterson asserted: "Advertising is becoming an art and the advertisements in the best publications are an embellishment. Without advertising the splendid periodicals of to-day would be impossible and readers are the ones most vitally interested. . . . The advertiser and publisher bring the reader in touch with the latest and best of everything in every nook and corner of the country. That is why you, as a reader, ought to be interested."[6] Patterson's terms here were similar to the *St. Nicholas* contest's request that readers help the magazine as consumers. The reasoning went that, first, readers benefited from the ads, since advertising support of the magazine made the magazine cheaper and better—that is, able to pay more for fiction and features. Readers ought therefore to reciprocate by showing their interest in the advertising. But, second, "the latest and best of everything" that the reader was brought in touch with evidently included the advertising itself, and not just what the editors could buy with its proceeds. Advertising provided particularly democratic delights, according to an advertising trade journal commentator, an "inviting path into the pleasures of the imagination, where all classes may revel in innocent delight."[7] As a second advertising trade journal writer put it, advertising "is news of the vital sort, news that shows the progress of the world up to date; while the literary stuff may have been lying for months in the editor's desk waiting for a chance to appear."[8]

The editor of *Good Housekeeping* made the same point directly to the consumer readership. Not only were ads good for them, but also the "Editor acknowledges that as readers 'discover' the advertising section of the magazine, his portion will have a formidable (though friendly) rival." These advertisements announced "only guaranteed merchandise and [were] written and illustrated by the most highly paid talent." *Good Housekeeping's* claim that readers found the ads valuable was followed by an orchestrated presentation of letters from readers (signed with initials). Their endorsements extended beyond merchandise itself in *Good Housekeeping's* ads, to praise even the "psychological influence emanating from clever advertisements" that caused the letter writer to want the new goods and to feel dissatisfied with her present possessions.[9] These letters presented other readers with a model of how to approach and use the advertising in the magazine; like the contests, these letters appealed to the eyes of advertisers, reassuring them that readers welcomed the ads into their homes and even sought them out.

Other ad-dependent magazines joined the chorus asserting that ads offered valuable entertainment and information. As if anticipating *Current Literature's* complaint that ads rendered the magazine disposable, Frank Munsey, in a *Munsey's* magazine editorial, hailed precisely the ephemeral quality of the magazine as what made it and its ads valuable: the magazine was timely, not ageless. Answering a reader who complained of the advertising bulk of *Munsey's* and who expressed the hope that "the day will come when every publication that we buy will be free from advertisements," Munsey argued that the advertiser is "a public benefactor. . . . It is through him that the reader keeps in touch with progress, with the trend of prices, with inventions and improvements. . . . In this age of invention, of mechanical perfecting, you are sure to wake up and find that you have bought something that is out of date, unless you watch the advertising pages of the magazine—that mirror of commercial enterprise."[10] It was only by keeping up with "news that shows the progress of the world up to date" that one avoided being dragged down into the pit of obsolescence. Advertising became the news of modernity.

Worth and Cost

Even if ads were as valuable themselves as these commentators claimed, they also had the role of supporting the literary mission of the magazine. A question emerged: Was the reader accepting an unwanted pile of ads in exchange for a lowered price for the literary matter of the magazine? Or was the reader being bribed by entertainment to read ads? By what calculus could the value of the reader's time be determined? Could the concept, hardly flattering to advertisers, that in buying a cheaper, ad-filled magazine, readers participated in an exchange in which paying attention to ads was a supplement to the price of the magazine, be shifted to the more ad-favorable idea that publishers simply filled readers' demands for ads? Commentaries and persuasive essays addressed these questions both in general periodicals and in the advertising trade press.

One strand of discourse pointed to the price paid by the advertiser for the ad: if it was worth so much to the advertiser, that should also be its worth to readers. This claim appeared just as editors and publishers began to publicize the amounts paid to well-known authors for their works, suggesting that the stories were provided as a generous gift to readers. And if an ad writer was paid so much money for an advertisement, then perhaps it was *more* worthwhile reading than a story for which a writer was paid less; at the very least it must be worth the reader's time to read it. Ads, fiction, and articles were placed on the same footing, measured by the same issue of pay. As Frank Munsey put it, "the advertiser . . . spends hundreds of millions of dollars annually to tell the reader—the wide-awake reader—just what he wants to know—what he should know."[11]

Samuel Hopkins Adams joined the question of payment to the ad writer with the cost of placing the ad. In his analysis, the words themselves in the ad become precious, and by extension take on value that they could not have in a literary work:

> Advertising is, in its best development, literature of the most expert and technical, though not of the highest, type. . . . In any extensive advertising campaign, letters of the alphabet are more expensive than pearls, and words than diamonds. The composer of a "national" advertisement is dealing with words which cost his principal at the very least a hundred dollars each, very possibly ten thousand dollars each. It behooves him to say the very most that can be said, to say it with the highest degree of explicitness, to appeal, to make it arrest, to make it convince, and all within the briefest possible compass. If that isn't literary work, I don't know what is. To a certain man having this quality highly developed, a salary equal to that of the President of the United States is paid. He earns it. Few men in the "legitimate" arena of literature earn or make more. Few do their "stunt" so well.[12]

For Adams, because the cost of the ad to the advertiser could be measured in a specific amount of money, that same sum must be carried over to the reader's accounting ledgers as its value to the reader. The reader would be foolish to discard such a gift: "The man who confines himself to the 'reading matter' of a modern high-class magazine is getting only part of what he pays for. The best experts of the day are striving, in a hundred phases of endeavor, to find something that will attract and amuse him, and he flings their work into the scrapbasket without so much as looking to see whether it hasn't something to say to him."[13] As *Collier's* prominent muckraking journalist, Adams placed the progressive reformer's imprimatur on the advertising in the "modern high-class magazine," where, he suggested, the male magazine reader, at least, should rely on the work of experts, presumably reliable professional men like himself.

Elsewhere, the juxtaposition of fiction and ad in such assertions of value became a recurrent joke, with commentators reviewing ads as though they were stories:

> [In] our old friend *Harper's Monthly* . . . there is a fantastic romance of a reduced gas bill, showing imagination of a high order, and some delightful examples of the eternal juvenile. . . . The editor of *Scribner's* is peculiarly lavish to his readers, giving them besides all this interesting and valuable matter (the variety of which can only be hinted at), a number of original illustrated jokes. . . . *McClure's* really gives us a great deal at very little outlay. There is a new and original study of Gold Medal Flour.[14]

Art in Advertising similarly reviewed magazine ads as though they were literature: "We notice a number of pages entitled 'Heating Apparatus,' which the editor has evidently been collecting with great care; a good many kinds of furnaces are referred to, all in terms of highest praise."[15] (Both pieces parodied the columns in which magazines and newspapers reviewed the month's magazines—columns that were themselves criticized as puffs.[16])

Other writings yoked ads and fiction by presenting a naïve character who prefers the ads to fiction, like Finley Peter Dunne's bartender, Mr. Dooley. Mr. Dooley begins with a simple inversion, assuming that magazines charge poets and writers to print their works, and are therefore growing rich at the expense of his favorite reading matter, which is being crowded out: "What I object to is whin I pay ten or fifteen cents f'r a magazine expectin' to spind me avenin' improvin' me mind with th' latest thoughts in advertisin' to find more thin a quarther iv th' whole book devoted to lithrachoor. . . . A man don't want to dodge around through almost impenethrable pomes an' reform articles to find a pair iv suspinders or a shavin' soap." Mr. Dooley further complains that ads and fiction are indistinguishable:

> Th' magazines ought to be compelled to mark all lithrachoor plainly so that th' reader can't be deceived. They ought to put two stars on th' end iv it or mark it "Reading Matter"; . . . As it is now manny iv these articles will fool nine men out iv ten. Ye pick up a magazine an' ye see something that looks like an' advertisement. It is almost as well printed an' illusthrated. On'y an expert cud tell th' diff'rence at th' first glance. But whin ye get to th' end ye find to ye'er disgust that ye've been wastin' ye'er time readin' a warruk iv fiction. It's very annoyin'.[17]

Mr. Dooley is angry to find that stories *begin* with oatmeal or somebody's extract or corned beef, or Cottolene, but end abruptly as fiction. He would evidently, with Earnest Elmo Calkins's ideal reader in chapter 3, prefer something that tells a business story all the way through. He would agree with another advertising industry commentator's assertion that "the public watches eagerly for each new chapter of the 'continued story' told in the advertising columns of well-known business houses."[18] Mr. Dooley goes on to praise the increase in advertising pages per issue in magazines, thereby echoing another advertising trade writer who called it "but a poor sort [of 'high-class magazine'] that cannot give its readers one-third more pages of advertising than of literary matter."[19] Mr. Dooley then performs the familiar maneuver of reviewing ads as though they were stories.

Dunne's satire played with the blurring edges of fiction and advertising within the magazines. The concepts that Dunne advanced satirically were not far from advertising promoters' hopeful claims about the value of ads: that ads were so worthwhile that readers sought them out and preferred magazines with many ads.

The discussion of the value of ads in the magazine took a more complicated course when it joined the question of what gender ads addressed and the place or status of the writer who addressed ad readers.

Gender and Magazine Reading

Who read these magazines and their ads, and where did they read them?

> Toward luncheon time Mrs. Doe has time to sit down for a half hour's rest and reading. Conspicuous on the living-table [*sic*] are the "Quarterly Review of the World," and a woman's magazine. Which one does she pick up? If she were a professor in a girls' college, the chances might be about equal, but being a home-manager instead, the "Quarterly Review" is "left at the post." Will she just turn to fashion pages, or suggestions for children, or picnic dishes or to a continued story? We do not know, it depends on her problem or mood of the moment. But this we do know—that all her common interests have a place somewhere in this magazine—and that it is to her the one "most popular book of the month."[20]

In anecdotes and graphics from the turn of the century, men reading magazines are often shown to be engaged in a purposeful pursuit. Magazine reading was seen as a form of intelligence gathering for men on the move; graphics portray men reading magazines on trains and streetcars, moving rapidly through the magazine as they move through the world; men wanted to "read as they run."[21] But for women, magazine reading was more often depicted as a pleasurable indulgence. Graphics show women reading in gracious homes or picturesquely leaning against trees; anecdotes discuss their reading as a leisure activity. Women evidently read magazines as a break from work. This division, though not absolute, appears in the popular advertising posters for magazines of the 1890s. And images of women reading both books and magazines at leisure outdoors in pastoral idleness were used to advertise a variety of products.[22] Such imagery is deployed, for example, in an 1891 advertisement for a hammock offered as a premium for securing subscriptions to the *Ladies' Home Journal*, showing both the hammock in use and the subscription seller resting by reading the *Journal* (see figure 6-1).[23]

Although these representations cast magazine reading as a leisure activity for women, according to another line of description, the magazines were the *trade journals* of the housewife trade, a phrase that the *Ladies' Home Journal* took up in its promotions to advertisers. The housewife "reads her own trade journal, in which she studies the market for the purchase of her raw materials, and learns the alchemy of her cooking. . . . From it she learns to buy what [other women] buy, to provide what they provide."[24] Women, in accounts of reading magazines, therefore commonly turned to the advertising section first, where ads carried information on the shopping that was an increased part of their daily work.

Nathaniel Fowler asserted in his compendium of advertising advice that although both men and women might "read magazine advertisements before they read the rest of the magazine, partly because the advertising pages are always cut, and largely because they may be more worth reading," this habit extended particularly to women, even in reading newspapers, where all the pages would have been cut apart, ready to read.[25] As one commentator wrote in 1909: "The housewife wishes to know what particular bargains are offered for the day of week. To her the department store space [in the newspapers] fairly teems with news, often news that she reads extensively before going back to the first page."[26] Newspaper ads, too, were overtly framed as trade press, with women as alert managers keeping abreast of the latest developments in the field: "The growing popularity of evening papers as advertising mediums also points to the fact that woman is the chief buyer. During the day's housekeeping she has experienced certain benefits by the use of articles brought to her notice by advertising, and when the work is done, it is an evening, and not a morning, paper she scans for new announcements."[27]

The assertion that women were particularly interested in the ads was sometimes raised as a saving answer to the question of whether *anyone* was in fact reading magazine ads. An advertising trade journal quoted an overheard conversation among "several famous advertisers" discussing whether the sixty to

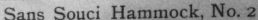

Figure 6-1 Reading magazines as a leisure activity for women. Ad for hammock given as a subscription premium, *Ladies Home Journal*, May 1891. (Courtesy of the General Research Division, The New York Public Library, Astor, Lenox and Tilden Foundations.)

one-hundred pages of advertising per magazine issue "is a drawback or an advantage":

> One of the gentlemen said he was convinced that there was a great number of people who always looked at this department first. "Did you ever notice," he said, "how a woman in the cars will read a magazine? She will take it up, look it all over carefully to see she has the real thing, read the baking-powder advertisement on the back, start in from the end, look all through the advertising pages—stopping now and then to read some announcement of special allurement—until she strikes the text. She then gives a sigh, reaches toward her back hair for a superfluous pin, and begins to mutilate the pages in a languid quest for the month's poetry."[28]

According to these advertisers' observation or wishful thinking, women read ads even in the general magazines, with their separate ad sections, with the systematic, ordered attention appropriate to business—the right approach to one's trade press—but also in pleasurable response to their allure.

These two ideas—that when women read magazines they engaged in a leisure activity, and that the magazine, not just women's magazines, but any magazine, was a trade press for women—seem contradictory. The claim that ads themselves were pleasurable for women, that shopping and choosing goods for the family was recreation, became the ground on which these seemingly contradictory ideas met.

Advertising Women

As we saw in chapter 5, within the stories in the middle-class magazines, marriage and the consumption it entails were antithetical to women's writing or other outside work. The issue of the woman writer was a weighty one in the middle-class magazines because the issue of the woman reader was a weighty one for advertisers. Unless middle-class women were full-time houseworkers, they would lack the proper orientation to the magazine. Women's approach to periodical reading—reading the ads before the articles and treating the ads as the news of their trade as housewives—was mocked, even in advertising trade writing, as foolish. And yet these habits were essential to the advertiser. Although throughout the 1890s magazine ads, even for such items as scouring soap, often continued to assume that men might buy them, while others addressed both men and women, women increasingly became the proper target of advertisements. Advertisers and magazine advertising departments began, albeit fitfully, to view women as the actual purchasers of household goods and realized that it made sense to address ads directly to them.[29] "The enormous increase in 'consumption goods,'" wrote one commentator, "is due more to women than to men. The bulk of modern advertising appeals more to the general household, and to the relief of housekeeping duties by means of novelties and contrivances, than to the business man."[30] Women were to be shoppers, while men were not only merchants, as they had been for decades, but advertisers and writers.[31]

As consumption was increasingly seen as gendered, components of consumption transactions were increasingly framed as gendered as well, playing out a scenario of man as the forceful seller and woman as the receptive buyer. As idealized in advertising trade discourse, the advertiser became the solid, worldly provider, and the woman a rather more pliable home-maker. In the version of this convention the *Ladies' Home Journal*'s parent company presented to potential advertisers in its booklet *Selling Forces*, "'Advertising,' said a speaker at a great business men's dinner, . . . 'makes for better furnished homes, better dressed people, purer food, better health, greater comfort, bigger life; and incidentally, advertising makes the advertiser a bigger, broader man—a national figure.'"[32] By 1905, an editorial cartoon in the advertising trade journal *Profitable Advertising* offhandedly drew on a gendered convention for rendering the relationship between the advertiser and the advertiser's audience: Cupid, labeled "the advertiser," shoots an arrow with the message "Won't you be my valentine" at a screen with a heart on it. A matronly woman labeled "the buying public" peeps out from behind the screen[33] (see figure 6-2). While Cupid presumably woos on behalf of the manufacturer or merchant, he is the most active player in the scenario. In Victorian England, the phrase "advertising man" referred to a man who created advertising, while "advertising woman" meant a woman who consumed the ads.[34] Although the latter phrase was not common in the United States, the opposition between male advertising creator and female advertising reader was posed as rigidly. Nathaniel Fowler put this idea in extreme form: "The woman who will not read advertisements is not a woman—consequently all women read advertisements."[35]

There was nothing new in framing buying and selling as a gendered relationship. The shift was into an emphasis on the gender relation and into including a larger cast of characters in it. What seems new as well is the uneasiness and irritation the newer players articulated at being positioned as suitors to women. A tension in advertising commentary underscored the gendered terms of the advertising/consumption transaction. Advertisers and editors were obligated to address women, and they resented what they saw as women's power over them in their role as shoppers. While ads flattered women as capable choosers, able to draw fine distinctions between items, commentary directed toward *advertisers* increasingly treated women derisively for being too concerned with trivialities like choosing between items. An anecdote in *Selling Forces* exemplified this tension:

> An impossible tale is told of a woman who entered the post office and asked, "Do you keep stamps?" The clerk produced the familiar large sheet of 100. She reached for the whole sheet, scanned it critically, and, placing her finger on one stamp near the center, said sweetly, "I'll take this one."
>
> This trait of fine discrimination in merchandise thus satirized accounts for the early and continuous success of advertising which is directed at women.[36]

Like a good speaker warming up an audience with a joke that raises and defuses its anxieties, the *Ladies' Home Journal* publisher reminded advertisers how capricious they found the women consumers upon whom their sales depended. Despite the disavowal of the tale as "impossible," what breaks through here is fear of the "fine discrimination" that advertisements and the magazine as a whole have nurtured in readers. If an advertiser's ads have encouraged consumers to see some fine point of superiority of their product over another, won't the consumer more readily see some even more infinitesimal difference pointed out by another advertiser? The publisher went on to reassure advertisers that approaching women through the *Ladies' Home Journal* would allow advertisers to harness this "trait of fine discrimination."

The derision directed toward the capricious and too-discriminating woman shopper responded to the inherent uncertainty of the advertising enterprise. When advertisers promoted a new or previously unadvertised product category,

HIS VALENTINE.

Figure 6-2   "His Valentine": editorial cartoon. (*Profitable Advertising* (February 1905, p. 975).

like bicycles, sales nearly inevitably increased. But as advertised products increasingly competed with other advertised products in categories like cereals, results were not so clear-cut. Even if their own sales increased, other companies might be even more successful. Advertisers became aware that they could not depend on any straightforward effort-response model; responses to their pitches were unpredictable. (Results of different ads were rarely measured until the 1910s.) This increasing emphasis on competition coincided with the growing recognition that women were the primary buyers for many products; the disturbing irrationality of the consumer at large was quickly translated into already existing stereotypes of the capricious woman. The uncertainty of marketing a product was blamed on women, even as ads and stories celebrated women's talents at consumption, shopping, and fine discrimination in choosing.

Advertising trade press and advertising advice books conducted an uneasy, complex discussion of what it meant to address women in advertising. As shopping became "a new feminine leisure activity," Rachel Bowlby has suggested, organized sales efforts by producers addressed to consumers thus "readily fitted into the available ideological paradigm of a seduction of women by men, in which women would be addressed as yielding objects to the powerful male subject forming, and informing them of, their desires."[37]And yet while the language of courtship and seduction appeared repeatedly in the advertisers' talk among themselves, as in such items as the Cupid and consumer cartoon, courtship requires ingenuity, and burdened admen with uncertainty. Women might not be sufficiently yielding; on the other hand, if they yielded too readily, what challenge or accomplishment would there be in wooing them?

The courtship paradigm was a powerful one for advertisers, it would seem, since courtship stories often appeared as sample stories for the advertising contests, in which shopping was often the chief activity. The female characters in such stories, as we saw in chapter 2, were wooed not just by human suitors but by a barrage of products and the advertising characters that represented those products.

The magazine courtship stories discussed in chapter 5 flattered women as choosers—as able to shop wisely for a husband as to choose among products. The converse of this image, articulated in the advertising trade press, finds anxious male advertisers and ad writers courting women who hold men's fate in their hands. When fiction appeared in advertising trade journals it often featured an adman in love, bringing his advertising skills to bear on courtship. In James Barrett Kirk's "A Midsummer Madness" in *Profitable Advertising* (1899), when Dick Walters's Daisy scorns him as insufficiently bold and venturesome, he persuades a company to let him take over its advertising for a month and saturates the August magazines with Daisy's photograph, captioned simply, "Mrs. Walters." While she is at first indignant to find that she cannot "walk down Broadway without being scrutinized by nine-tenths of the people she met, while many times she overheard the whispered speculation: 'That must be Mrs. Walters,'" she admires Dick Walters's "daring spirit."[38]

As a result of the ad, she hears herself spoken of as Mrs. Walters, and both tries on that role and enters into the adman's fantasy. The ordinary reader must make an imaginative leap, both to identify with the figure in an ad and to try on the position constructed by the ad, but Daisy, like the scrapbook compilers who disregarded questions of resemblance and jumped straight to identification, finds herself already represented within the ad; the question for her is whether to accept the role in which she is cast. The adman is sufficiently bold; when she indignantly asks why he placed such ads, he answers, "Because I wished it were true," and obtains her hand. Just as a shopper accedes to the advertiser's wish by purchasing the product and thus matching more closely the figure in an ad using the product, Daisy typologically fulfills the picture of Mrs. Walters, becoming the picture by becoming Mrs. Walters. The story further links courtship and advertising as Dick explains, "there are two things in which audacity is absolutely necessary. . . . Advertising and—love." The story ends with a reference to Dick's "framed photograph of the ad girl, who seemed to smile approval of his audacity" from its trophy-like position on the wall.[39] "A Midsummer Madness" presents the female ad reader as an ideal ad recipient who pliantly remakes herself to fit the role the adman presents to her. The story reassures admen that advertisements *can* succeed in wooing demanding and capricious women.

Gendered Ad Writing

When magazine essays praising advertising served the double function of training the readers' attention on ads and reassuring advertisers that their ads were read, they made an implied promise to the male advertiser that he could present his ads to an admiring female audience. But in discussions of advertising within the advertising profession, admen expressed resentment of this power of choice that women evidently possessed. The construction of a gendered language of advertising, with different writing styles addressed to women and men, was in part an expression of this resentment, in that the style said to be suitable for men (and generically suitable for ads) was also the more concise, succinct style being increasingly praised in fiction as *modern*. The style said to be suitable for women, on the other hand, was increasingly devalued as trivial, old-fashioned, and overwritten. Within this paradigm, to court women shoppers, male ad writers ostensibly had to demean themselves by twisting their prose into a retrograde knot.

The notion among advertisers that advertising was valuable and appealing to everyone, and therefore worthwhile reading matter, collided with the belief that women were more devoted readers of advertising than men. Advertisers were consistently advised that ads written for men would have to be brisk, since men wouldn't waste time reading ads: "In appealing to the trade of men, short concise statements of fact are the most effective: men want to read as they run—and will not wade through the descriptive advertising that appeals to women."[40] To address ads to men, who want "a quick story interestingly told," an adman should simply be himself:

> Be perfectly natural, be vigorous when you feel like it, be easy when you feel like it. Speak your own thoughts. Be true to yourself in this regard. The more a man writes advertising the more confidence he has in the power of his pen and the more individuality and consequently interest will his advertising possess.[41]

> Men are reached by a strong swift style—a style the reflex of their business life—a style that does not lose itself in a maze of details and wanderings into fashionable features—a style sententious, business-like, pleasant and at times a trifle humorous.[42]

But men writing ads would have to contort themselves to address women: "Women, on the other hand, are reached with an easier and more detailed style—a style which never loses sight of the value end of the article advertised while showing no hesitancy in going into deep details about material, brand, manufacture, color, shade and every little point about the article advertised." Ads written to women could be long-winded, since women would enjoy lingering and fantasizing over the particulars. Writers addressing women would have to learn to emphasize "details" and "trifles" and to train themselves to point out the connection to style and to "the devious pathway of the will-o'-the-wisp Fashion—whose imperious mandates are blindly followed by every woman."[43]

The idea that women would read long descriptive copy for which men have no patience echoes the pairing of men who read magazines on the fly or at desks while women read lounging in hammocks or easy chairs; it appears repeatedly (see figure 6-3). So a 1904 article in *Profitable Advertising* advised advertisers to distinguish between the urban male reader on the one hand, and the rural male reader and women readers in general, on the other. It complained that in writing direct mail solicitations, companies too often failed to vary letter lengths according to audience: "There is no attempt to make the letter short for the busy business man, but long for the farmer or woman who has plenty of time and wishes to have a letter that will require time to read."[44] A solicitation that imposed on the valuable time of the busy businessman supposedly provided pleasure and entertainment for women. This may have been because housewives' time had no market value and, like farmers' time, was not measured off in hourly wages or salary.[45] Despite the concurrent characterization of magazines as women's trade press, ad reading was defined within this paradigm as a *work* activity for urban men, in which evaluating the product, like other forms of labor, should be done efficiently. Whether they briskly evaluate a proposed office purchase or productively absorb information from the magazine, shopping was not an activity for men to linger over. For women (and farmers), however, ad reading was classed as a form of entertainment, a category in which one would ordinarily pay to have one's time occupied.[46] Circling back to the Barnumesque idea of ads as entertainment for the naïve, ads were cast as a source of pleasure to women, although by this point they were expected to be non-Barnumesque progressive advertisements, telling the "business story."

Advertisers and ad trade journals suggested that women felt a special need for ads. In 1903, a former editor of the advertising trade weekly *Printer's Ink* made what had become a standard claim, that advertising was educational, but asserted that women were in special need of such education: "If a woman did not read advertisements she would not have learned of the new crackers, breads, canned stuffs, and a host of other recent delicacies. . . . It is almost impossible to conceive what an ignorant creature the modern woman would become, without the aid of the advertisement of to-day." Advertisements in the monthly magazines are good for everyone, he says, but "the influence of these advertisements is vastly greater on a woman than on a man. She is more readily convinced, and, moreover, has at her command a far greater amount of time to read, and consider, and inquire."[47]

A suggestive contradiction emerges in the instructions and commentary on styles of ad writing. Commentators cast women as the buyers, the main target

Figure 6-3 A somewhat less judgmental version of the idea that ads should be written differently for men and women appears in this advertising agent's notice in an advertising trade journal. The insets show a man reading his magazine at a desk, while a woman reads at leisure. (*Profitable Advertising*, September 1900, p. 229.)

to address in most advertising. They asserted that women didn't simply put up with, but actually enjoyed reading long-winded ads, and that a special style was needed to address them. At the same time, these advisors and commentators formulated generic rules for advertising writing: ads should be short and succinct; every word should count. Instructions for writing ads in general, then, matched those for writing ads to men, and yet women were said to be the main ad readers.

This seeming paradox could simply reflect the conventional notion that an audience that is not specifically gendered is male. But it points as well to the relative prestige of addressing men rather than women. The best advertising writers, one advertising trade journal commentator explained, were newspaper reporters, because they have acquired "a terse, vigorous style that experts deem wholly essential in advertising, as every word used therein represents so much outlay of money and appeals instantly to commercial sense."[48] The linguistic economy here parallels Samuel Hopkins Adams's assertion that the cost of the words to the advertiser should make them more valuable, and in this case appealing, to the reader. Men were understood to be the real readers of newspaper journalism, since women, "after reading the birth, marriages, and deaths, run through the advertisement columns and then throw the paper down as exhausted."[49]

This "business" style was increasingly identified with the kind of consice "modern" writing in demand for magazine fiction. It was cast as the antithesis both of material addressed *to* women and written *by* women. Frank Munsey, editor and publisher of *Munsey's*, for example, complained repeatedly in his editorial columns about the kinds of "thin, flabby, inane" stories the magazine received,[50] identifying them specifically as women's writing:

> Of the something like three thousand manuscripts that come to us from month to month, ninety to ninety-five percent come from women. It is not our purpose to discourage women contributors, but to encourage men contributors. We have said before that we want force in our fiction—virility, action, plot, clever handling. . . .
>
> The utter weakness of the average alleged story is appalling—sickening even. . . . It is a process of building around nothing and labelling it fiction. It has occurred to us that possibly women are more apt to fall into this error than men. This is why we want to encourage men to take up literary work—men with force, strong, manly men, men with imagination, with depth of human nature.[51]

He called "for us as a nation" to "begin to do men's work in fiction, as we are doing it in the industries of the world."[52]

The "strenuous and masculine" style was a hallmark of the Progressive era, Christopher Wilson notes in his study of literary professionalism in this period. Or at least applying such terms to favored writing was. The "masculine ethos interlocked with other 'realistic' keywords," so that Walter Hines Page, of the book publishers Doubleday and Page, for example, asserted his preference for the "roast beef" of literature—a quintessentially men's dish—while the editors of *Cosmopolitan* promoted one writer's work as "vascular and virile."[53]

Frank Munsey hailed the concise "masculine" style as well, as "men's work in fiction," which would be like advertising: "The modern advertisement is a thing of art, a poem, a sledge hammer, an argument—a whole volume compressed into a sentence. Some of the cleverest writing—the most painstaking, subtle work turned out by literary men today—can be found in the advertising pages of a first rate magazine. . . . The man of original ideas and keenest pen tells his story in a word."[54] (Munsey perhaps anticipated Mr. Dooley's praise of advertising's improving effect on the tone of literature by "givin' lithry men a good wurrukin' model to follow.") The "masculine" writing Munsey praises here is very different from that of the writers Munsey dislikes, "who fancy that a story cannot be told without high flown periods and ornate phrases."[55]

These "ornate phrases" sound not only like the story by Elizabeth Hastings Pratt that the fictional editors reject in Adelaide Rouse's "Story of a Story" discussed in chapter 4, but also like the copywriting style Helen Woodward took up around 1907 when she switched to writing ads directed to women. The title of this chapter of her memoir, "Translated into the Feminine," emphasized the idea that men and women spoke, or at least read, different languages: "My early advertising about books was bullet-like in directness and addressed entirely to men. It was raucous, over-emphatic, and full of superlatives. . . . [In writing for women] I wrote the sample advertisement exactly in the same manner as the book copy. Then sentence by sentence, word by word, I softened it, sweetened it, made it sticky with sugar. Whenever I thought it was needed, I stuck in a *cute* word."[56]

Ultimately, the paradox involved in tailoring levels of conciseness for male, female, and generic readers was resolved with the claim that *everyone* preferred concise writing; the ground of distinction between men's and women's ads shifted to other stylistic criteria. Women, says Woodward, "like to be approached softly and cleverly, they really demand even more definiteness than men. They want facts, provided the facts are stated simply and gently."

Ellis Parker Butler's Perkins of Portland story "The Adventure of the Poet" comes to a similar conclusion. Perkins's business partner, the unnamed narrator, competes with a poet for the love of Kate. He has Biggs, his staff jingle writer, compose poetry for him, with lines like

> Your lips are like Lowney's
>   Bonbons, they're so sweet;
> Your eyes shine like pans
>   That Pearline has made neat.[57]

Kate and her poet suitor are amused by these attempts, which include a poem praising Kate for reading magazine advertising pages first:

> She feels the pulse-beats of the world of men,
>   The power of the advertiser's pen.[58]

But in a partial reversal of the courtship paradigm discussed in chapter 5, where women reject richer suitors for poor but striving ones, Kate turns out to have a more directly mercenary approach. The poem that finally succeeds in wooing her explains in detail just how profitable is the narrator's partnership in Perkins

& Co. The story thus collapses the selling triangle into a couple: Kate does not buy the ad—either Biggs's creation, or the poet-suitor's—but rather the product, the affluent narrator. Butler's joke is that Kate claims she wants to be wooed by poetry but is finally swayed by cold facts.

Summary

By 1913, the *Ladies' Home Journal*'s publisher was willing to claim in its booklet *Selling Forces* that the ads in its own magazines were superior to the fiction:

> Advertising is news—news for the housekeeper and the home owner, for the man who wants an automobile, for the woman who is buying new clothing or any one of a hundred daily necessities and luxuries. It is up to the minute. The leading fiction may deal with the Court of King Arthur; the back cover advertisement must deal with *tomorrow*. It has been said that "some folks find the cereal advertisements as interesting as the serial stories."

It inadvertently made the claim a more general one by quoting from a writer in *Good Housekeeping*:

> Anne O'Hagan makes one of her characters say: . . . "I receive whole, copious courses of instruction in household arts merely by 'writing for booklet.' I learned how to take off varnish, how to paint, enamel, stain and finish woodwork, and how to do over furniture by writing to the paint makers for their booklets. Take advertisements seriously? Of course I do. It's the only way for any woman to take them who ever buys anything."[59]

Here, as well as in the earlier scenarios in which women turn to the advertisements first, magazine publishers state explicitly that women find something especially necessary and worthwhile in the ads. But what happens after the hairpin is brought into play, the pages cut apart, and the editorial or reading matter opened up? Within this scenario, women read the editorial matter and the advertising with similar attention. This alleged lack of discrimination was a focus of mockery from the 1890s through the 1910s: although advertisers busily insisted on the high value of their ads to the reader, they were disconcerted when women seemed too readily to agree with them. Women seemed not to know what was valuable in the magazine, unable to tell the difference between ads and editorial matter or between ads and fiction. This evident refusal to see boundaries expressed what was going on in the magazines themselves: ads spoke for individual advertisers, while the editorial matter of the magazines spoke for the interests of the advertisers in the aggregate.

The women's magazines and household publications were the first monthlies to mix ads with editorial matter, normalizing that practice for the mass magazines that soon followed suit. The assertion that women readers in particular found ads valuable, and that these ads were crucial to their work, made the mix of ads, fiction, high-art ad references, and uplifting essays on a page no longer a reflection of carnivalesque disorder or intolerable intrusion. Ads were far from being zoned out of the magazines; they were virtually mandated.

Conclusion:
Technology and Fiction

When you bind your magazines by all means bind in the advertising pages. If it makes too bulky a volume leave out some of the literature, for to future generations this will be of the least importance.

editorial, The Independent, 1902[1]

Magazines are not the principal national mass medium they were at the turn of the century; television long ago overtook them. The consumer culture of the 1890s contained only a faint foreshadowing of the pervasive and complex interplay between fictional narrative and advertising that exists today on television. But TV commercials have also continued to sprawl into written fiction in ways more familiar from the 1890s.

In 1993, for example, a London publisher brought out a novel based on a series of television commercials for Nestlé's Taster's Choice coffee (called Gold Blend in England); it landed on England's best seller list. The book's conception was more like an ad's than a novel's: an editor at Corgi books came up with the idea and commissioned an already successful author to write it under a pseudonym. The front and back covers carried the Nestlé logo and pitched the book to readers who knew the commercial.

Like the turn-of-the-century contests that invited readers to take advertising characters beyond the ads and to enjoy their company—in both senses—*Love Over Gold* invited viewers-turned-readers to follow the characters of what the book cover billed as "TV's greatest romance" beyond the lines of the commercial. The cover copy suggested, "If you think you know the story . . . try reading between the lines" to discover the "real" characters behind the commercial: "What do you really know about the Gold Blend couple?" Readers of the 343-page novel had already entered into the joke, by being willing to buy a book based on an ad, and the book capitalized on this further with the broad wink of an offer of a video of the whole set of commercials: "Enjoy your favourite commercials without a break."[2] The irony and self-parody apparent in some chromolithographed trade cards of the 1880s have been refined as a staple of print and especially television advertising, flattering the audience as people too smart to be taken in by commercials, who can therefore sit back and enjoy them.[3]

The characters, once firmly associated with the product, became the stand-in for the brand, so the novel itself need not mention the coffee to reinforce the commercial message. In fact, although the novel is fairly afloat with the consumption of liquids, Gold Blend is named by brand only twice. The far more frequent scenes of the characters drinking high-quality espresso or expounding in detail on their wine choices serve simply to align their taste in upscale beverages with their appreciation of the instant coffee of the sponsor.

In creating the TV commercial narrative that sold its products, Nestlé created a surplus of publicity, which Corgi could then appropriate for its book, thereby retrospectively causing the coffee ads to advertise its books. In this sense, this is how ads for soda also sell tee-shirts emblazoned with Coca-Cola logos. But in *Love Over Gold* the advertising constituted a narrative as well as an image, and it was one that viewers found compelling enough to want to participate in further. The compressed, teasing narrative of the commercials invited readers to enter into and fantasize about the scenario themselves, and then be interested in staying in it longer by reading what someone else did to flesh it out.

We've become accustomed to the flip side of the Taster's Choice move — product placement, or the seemingly casual display of products in movies, paid for by the product manufacturers. Broadway theater has even, for serious money, taken up Ellis Parker Butler's joking suggestion that plays incorporate product mentions, and it stocked a 1995 revival of *How to Succeed in Business without Really Trying* with references to Eight O'Clock Bean coffee, garnering cooperative publicity in payment. Such uses of product names and references have been naturalized by the unpaid mention of brand-name goods in fiction, a device popular with some writers as a way of signaling that the reader exists in the same world as the characters and as a shorthand way to frame the tastes and class of their characters. Following in the footsteps of Amelie Rives, with her 1888 use of Pears' soap in *The Quick or the Dead?*, such writers as Bobbie Ann Mason, Bret Easton Ellis, and Frederick Barthelme made product mentions hallmarks of American fiction in the 1970s and 1980s. TV commercials and their supercharged, compressed scenarios have undeniably exerted formal influence on modern and postmodern fiction, as well.

A century after the events and practices laid out in this book got their start, Americans have become accustomed to the interplay of advertising and fiction. Magazines without ads look a little odd. Magazines like *Consumer Reports*, *Ms.*, or *Adbusters*, which pointedly eschew ads, nod to the expectation that ads should be present even as they take advantage of their freedom from advertisers' control, by running dissections of ads, or parody ads.

IN THE LATE NINETEENTH century, American fiction, like the domestic household of the period, made contact with the new mass-produced brand-named merchandise. Proliferating mass-produced goods raised troubling questions about whether human relationships, as well as goods, might be reproducible. If what was essentially the same bar of soap—a bar of soap that had come from a single central point of manufacture and was identical to many other bars of soap with the same name and identical wrappers—could be in many households at once, then maybe those households were the same and reproducible in other respects. If each household were filled with goods identical to those in other households—if everyone in each household wore the same items and washed with the same scented soap—then maybe the people in those households were not so individual and distinct from one another.

Advertising, both a product of and promoter of new technologies, told its readers that unique subjectivity could be expressed through use of mass-produced goods: to bring out your *individual* beauty, buy our brand of soap, even though thousands of other women are buying it. Readers of advertising were encouraged to see mass-produced products as not just compatible with, but even the material for, constructing their unique individuality.

So it was not just the technologies of mass production, improved, cheap freight transportation, and packaging that allowed the large-scale manufacture and nationwide distribution of this new type of goods that inspired anxiety. Advertising pressed into new areas in the late nineteenth century, moving into what had been defined as private, noncommercial space. Advertisers wanted to make this move in part because they believed their appeals were more likely to be listened to in the parlor than in the store. As advertisers created a cultural shorthand that made brand-name references universally accessible, allusions to articles in use within the individual and private world of the home, even in intimate use, took on a new kind of public accessibility.

Manufacturers learned, in other words, to appropriate the technologies of fiction, as well as of transportation and mass manufacture, to sell more goods. And, as it turns out, by the late-twentieth-century, manufacturers' use of narrative techniques, united with wider media dissemination, has become so successful that readers are apparently eager to find out what brand of soap a happy couple will discuss next, and what happens after the advertised brand of coffee is sipped.

Rotes

Introduction

1. Edward Bok, *The Americanization of Edward Bok: The Autobiography of a Dutch Boy Fifty Years After* (New York: Scribner's, 1921), 292–293.

2. For more on department stores, see Elaine Abelson, *When Ladies Go A-Thieving: Middle-Class Shoplifters in the Victorian Department Store* (New York: Oxford University Press, 1989); Susan Porter Benson, *Counter Cultures: Saleswomen, Managers, and Customers in American Department Stores, 1890–1940* (Urbana and Chicago: University of Illinois Press, 1986); William Leach, *The Land of Desire: Merchants, Power, and the Rise of a New American Culture* (New York: Pantheon, 1993); Sheila M. Rothman, *Woman's Proper Place: A History of Changing Ideals and Practices, 1870 to the Present* (New York: Basic, 1978); and Rosalind Williams, *Dream Worlds: Mass Consumption in Late Nineteenth-Century France* (Berkeley: University of California Press, 1982).

3. The present-day designer Milton Glaser also notes similarities between magazines and supermarkets: "The concept that one moves through the pages of a magazine while being entertained and informed has something to do with the idea that one moves through aisles of a supermarket while also being entertained and informed. There are issues common to these different motions through space; for example: how does one engage people's attention?" Marshall Blonsky, *American Mythologies* (New York: Oxford, 1992), 218–219.

4. "Fiction and Advertising" in column "From 'P.A.'s' Point of View," *Profitable Advertising*, February 1905, p. 1089.

5. An excellent study that takes on a magazine in its entirety is Catherine Lutz and Jane Collins, *Reading National Geographic* (Chicago: University of Chicago Press, 1993).

6. I have borrowed Philip Fisher's comparison of the urban streetscape to the open floor plan of houses that appeared around 1910, creating the mode of outdoor life inside. His description could equally be applied to the turn-of-the-century magazine: "What the open floor plan encourages is a variety of actions and tasks taking place within sight of one another. While one person cooks, another reads, the children play, and each is aware of the others in an intermittent way. Each interrupts the other from time to time." Philip Fisher, *Hard Facts: Setting and Form in the American Novel* (New York: Oxford University Press, 1985), 138.

7. Simon J. Bronner's "Object Lessons: The Work of Ethnological Museums and

Collections," on the development of museum displays of artifacts, and William Leach's "Strategists of Display and the Production of Desire," note the similar display strategies used by museums and department stores of the period. Both in Simon J. Bronner, *Consuming Visions: Accumulation and Display of Goods in America 1880–1920* (New York: Norton, 1989).

8. Frank Luther Mott, for example, writes of the "ten-cent magazine revolution" in 1893. Mott, *A History of American Magazines, 1885–1905* (Cambridge: Belknap Press of Harvard University Press, 1957), 46.

9. The new magazines also followed the elite magazines in their sale and promotion of single issues of the magazine, rather than inviting readers to join a community of readers as the mail-order monthlies did through their emphasis on subscriptions and subscription clubs.

10. See Richard Ohmann's "Where Did Mass Culture Come From? The Case of the Magazines" and "Advertising and the New Discourse of Mass Culture," in his *Politics of Letters* (Middletown, Conn.: Wesleyan University Press, 1987). I am indebted throughout this book to the initial prompting of Ohmann's work.

11. Mott, *History of American Magazines*, 8.

12. A few other magazines occupied a middle ground. *Lippincott's*, for example, lowered its price to twenty cents a copy in the 1880s by eliminating illustrations. John Tebbel and Mary Ellen Zuckerman, *The Magazine in America: 1741–1990* (New York: Oxford University Press, 1991), 67.

13. Christopher Wilson, "The Rhetoric of Consumption: Mass-Market Magazines and the Demise of the Gentle Reader, 1880–1920," in Richard Wightman Fox and T. J. Jackson Lears, eds., *The Culture of Consumption: Critical Essays in American History, 1880–1980* (New York: Pantheon, 1983), 45. See also Christopher P. Wilson, *The Labor of Words: Literary Professionalism in the Progressive Era* (Athens: University of Georgia Press, 1985). I am indebted in this section to Wilson's analysis of the change in the editor's role.

14. The names appear in an attack on the elite magazine world for being ingrown, complaining of the reciprocal promotion of the writers of these three departments. "Our New Department," *Art in Advertising*, November 1890, p. 170.

15. L. Frank Tooker, *The Joys and Tribulations of an Editor* (New York: Century, 1923), 322–324.

16. Mott, *History of American Magazines*, 482.

17. Brander Matthews, *The Story of a Story and Other Stories* (New York: Harper's, 1893), 3–50.

18. Matthews is accurate in suggesting that the magazines had a secondary readership beyond the newsstand. Laura Ingalls Wilder, for example, mentions used magazines as an item that arrives in bundles from the east in the 1880s. Laura Ingalls Wilder, *Little Town on the Prairie* (1941; New York: Harper & Row, 1953), 215. Anne Ellis, growing up in a Colorado mining town, reported reading 1890s issues of "*The Cosmopolitan*, much dogeared by the time it got around to us," along with other publications at least ten years old. Other accounts report catching from "the flotsam of print that happened to drift to the farm . . . periodicals left by visiting neighbors," along with story papers and mail-order monthlies. The diary of a turn-of-the-century Georgia farm woman records her gratitude for a neighbor's stack of old *Ladies' Home Journals*. Anne Ellis, *The Life of an Ordinary Woman* (1929; Boston: Houghton Mifflin, 1990), 125; Mark Sullivan, *The Education of an American* (New York: Doubleday, Doran, 1938), 70; Charles LeGuin, ed., *A Home-Concealed Woman: The Diaries of Magnolia Wynn LeGuin* (Athens: University of Georgia Press, 1990).

More formal mechanisms for sharing magazine subscriptions also existed that point to how social a practice magazine reading could be. At least one group of readers formed a "Neighbors' Magazine Club" for sharing subscriptions to sixteen magazines among thirteen households. Participating households each received and sent on between four and seven magazines a week, including both the new ten-cent magazines *Munsey's*, *McClure's*, and *Cosmopolitan*; the household magazine *Ladies' Home Journal*; the elite magazines *Harper's Monthly*, *Scribner's*, and *The Century*; as well as the more specialized *International Studio*. This group of avid readers, in other words, did not confine itself to one or the other end of the elite/nonelite spectrum. From "1902 Neighbors' Magazine Club. Read Regulations," n.p. This cover sheet, attached to the December 1902 issue of *International Studio* is printed with club members' names and addresses and includes space to note the dates magazines were received and sent on, along with a list of regulations. From the collection of the author.

19. Ohmann's assertion—that like the elite magazines, the ten-cent magazines excluded first- and second-generation immigrant readers—is problematic, given that Samuel S. McClure, editor and publisher of *McClure's*, and Edward Bok, editor of the *Ladies' Home Journal*, were both immigrants themselves, albeit northern European ones. What is certainly true is that working-class immigrants from Eastern and Southern Europe or from Asia were not likely to have found their lives or concerns reflected in these magazines, and would have found people like themselves treated as exotics in them.

20. Mott, *History of American Magazines*, 543. Because the exclusion applied to new advertising contracts, not to advertising that was already under contract, issues of the magazine didn't immediately reflect the change.

21. *Journal* policies that regulated the appearance of the ads—by excluding certain uses of display type and demanding a hand in the design of the ads—although irksome to individual advertisers, similarly had the effect of making advertising in the magazine as a whole more appealing, and made it a more desirable place to advertise. See John Adams Thayer, *Astir: A Publisher's Life Story* (Boston: Small, Maynard, 1910), 81–87. Thayer was advertising manager of *Ladies' Home Journal* during this period.

22. Sullivan, *Education of an American*, 136.

23. Gloria Steinem, "Sex, Lies and Advertising," *Ms.*, July/August 1990, pp. 26, 27.

24. Frank Munsey Jr., "The Modern Advertisement," in column "The Publisher's Desk," *Munsey's*, July 1895, p. 438.

25. For a celebration of the pleasures of ballyhoo, see Chalmers Lowell Pancoast, *Trail Blazers of Advertising: Stories of the Romance and Adventure of the Old-Time Advertising Game* (New York: Frederick H. Hitchcock, 1926).

26. For more on the term "progressive," see Daniel T. Rodgers, "In Search of Progressivism," *Reviews in American History*, December 1982, pp. 113–132, and Richard Hofstadter, *The Age of Reform* (New York: Vintage, 1955).

27. For more on the development of national markets, see Susan Strasser, *Satisfaction Guaranteed: The Making of the American Mass Market* (New York: Pantheon, 1989). For a discussion of the interdependence of magazines and advertisers of nationally distributed goods, see Ohmann, "Where Did Mass Culture Come From?"

28. Bok, *Americanization of Edward Bok*, 292.

29. Amy Kaplan, *The Social Construction of American Realism* (Chicago: University of Chicago Press, 1988), 10.

30. William Dean Howells, *A Hazard of New Fortunes* (1890; New York: New American Library, 1965), 131.

31. Michael Schudson discusses ads as short-hand realism in *Advertising, the*

Uneasy Persuasion: Its Dubious Impact on American Society (New York: Basic, 1984), 157.

32. William Leiss, Stephen Kline, and Sut Jhally, *Social Communication in Advertising: Persons, Products, and Images of Well-Being* (New York: Methuen, 1986), 121. Leiss et al. draw on Merle Curti, "The Changing Concept of 'Human Nature' in the Literature of American Advertising," *Business History Review* 41, no. 4 (winter 1967): 335–353. Jackson Lears locates the roots of this shift around 1910, with the "Reason Why" approach's promise to the consumer of a "richer, fuller life" from using the product. Other advertising historians such as Stephen Fox and Susan Strasser each show advertising by atmosphere and association beginning to displace hard-sell copy around 1915, while Roland Marchand similarly finds advertisers before World War I only just beginning to shift to "selling the benefit instead of the product . . . prestige instead of automobiles, sex appeal instead of mere soap," focusing, in other words, on the pleasures and psychic advantages of ownership. See T. J. Jackson Lears, "From Salvation to Self-Realization: Advertising and the Therapeutic Roots of the Consumer Culture, 1880–1930" in Richard Wightman Fox and T. J. Jackson Lears, eds., *The Culture of Consumption: Critical Essays in American History 1880–1980* (New York: Pantheon, 1983), 18; Stephen Fox, *The Mirror Makers: A History of American Advertising and Its Creators* (New York: William Morrow, 1984), 70–72; Strasser, *Satisfaction Guaranteed*, 159–161; and Roland Marchand, *Advertising the American Dream: Making Way for Modernity 1920–1940* (Berkeley: University of California Press, 1986), 10.

33. Leiss et al., *Social Communication*, 124; Marchand, *Advertising the American Dream*, 105–108.

34. The magazine was safe space in another sense as well. Although each month's contents were different, the reader could be confident of what he or she would find in it. The reader knew in which department to look to find another installment of Henry Ward Beecher's sermons in *Ladies' Home Journal*, or a column of very short stories in *Munsey's* that was sure to include a courtship story, or the art reproductions in the same magazine that were sure to include bare bosoms. Michael Schudson has discussed how predictability and uniformity, along with the related quality of ease of use, became desirable attributes of products in this period. This same desire for reliability, for predictability, that appears in the concurrently developing demand for brand-named products and for nationally available products of uniform quality, helped to build continuity of monthly readership for the magazines.

Chapter 1. Readers Read Advertising into Their Lives: The Trade Card Scrapbook

1. Both quotes from Janet E. Ruutz-Rees, *Home Occupations* (New York: D. Appleton, 1883), 95, 96, 98.

2. Letter by Lucy Waterbury, in 1882 newspaper clipping from *The Helping Hand* preserved in diary jointly kept by Norman and Lucy Waterbury. The diary is in the Barbara Rusch Collection of 19th Century Advertising and Manuscript Ephemera in Thornhill, Ontario; I am indebted to Barbara Rusch for supplying me with information about and photocopies of the diary.

3. See Neil McKendrick, "Commercialization and the Economy," in McKendrick, John Brewer, and J. H. Plumb, *The Birth of a Consumer Society: The Commercialization of Eighteenth Century England* (London: Europa, 1982). McKendrick points to rising wages and the employment of women and children in industrial work as both making more income available for spending and creating a demand for goods the now-industrially

employed women would otherwise have made at home: "clothes, beer, candles instead of rush lights, manufactured cutlery and pottery instead of home made treen, furniture, etc." (23).

4. Although Susan Porter Benson uses the phrase "palaces of consumption" to refer to the even more architecturally elaborate stores that were remodeled or built after 1890, with a "double interest in luxury and efficiency" (82)—that is, joining the new emphasis on efficiency to the already existing emphasis on luxury and service—the phrase seems equally descriptive of the earlier stores from the point of view of customers, whom Elaine Abelson documents describing themselves as overwhelmed to the point of dizziness by the sumptuousness of the stores. Sheila Rothman sees the stores as "cathedral-like" and quotes the *New York World* calling Macy's "a bazaar, a museum, a hotel and a great fancy store all combined" (19). Susan Porter Benson, *Counter Cultures: Saleswomen, Managers, and Customers in American Department Stores, 1890–1940* (Urbana: University of Illinois Press, 1986); Elaine Abelson, *When Ladies Go A-Thieving: Middle-Class Shoplifters in the Victorian Department Store* (New York: Oxford University Press, 1989); and Sheila M. Rothman, *Woman's Proper Place: A History of Changing Ideals and Practices, 1870 to the Present* (New York: Basic, 1978). See also William Leach, *Land of Desire: Merchants, Power, and the Rise of a New American Culture* (New York: Pantheon, 1993).

5. Both of the main collections of scrapbooks I examined—in the Thelma Mendsen Collection at the Winterthur Library, Winterthur, Delaware, and at the Margaret Woodbury Strong Museum, Rochester, New York—were collected in the Northeast. Many more scrapbooks are in the hands of private collectors, and countless others have been dismembered by present-day dealers in trade cards, who soak the cards from the pages and sort them into the categories they currently use.

6. During this shift, agents became more involved in writing copy and ultimately in coordinating advertising campaigns. The advertising trade publications such as *Printer's Ink, Profitable Advertising*, and *Judicious Advertising and Advertising Experience*, which have also provided chroniclers of advertising with source material, were, similarly, published by early advertising agencies whose income relied on placing ads in newspapers and magazines and who wrote almost entirely about those media. These trade publications were directed both at people involved in advertising in newspapers and magazines and at people in manufacturing and retailing, whose duties included arranging for their companies' advertising.

7. Robert Jay, *The Trade Card in Nineteenth Century America* (Columbia: University of Missouri Press, 1987), 3.

8. Carte-de-visite photos of celebrities might also be purchased. Pinkham's card might thus alternatively position her as a prominent authority.

9. See Susan Strasser, *Satisfaction Guaranteed*, for more on the switch to brand names. Some of the cards for brand-named merchandise bear the name of a grocer or other storekeeper as well, and some were printed on the back with a message from the grocer, suggesting that the cards were supplied by the manufacturer with blanks for this purpose and that retailers themselves saw the trade card as sufficiently valuable to customers that they preferred using it to a blank card. Strasser points out that it was often not in the interest of the retailer to sell brand-named staples since the profit margin on them was usually smaller; the advertising was evidently an incentive. Susan Strasser, *Satisfaction Guaranteed: The Making of the American Mass Market* (New York: Pantheon, 1989).

10. Jay, *Trade Card*, 39–40.

11. Card for Lautz Bros. & Co., Master Soap. From the collection of the author.

12. For an excellent explanation of the lithographic process, see Peter Marzio, *The*

Democratic Art: Pictures for a Nineteenth Century America, Chromolithography 1840–1900 (Boston: Godine, 1979).

13. Nathaniel Fowler Jr., *Fowler's Publicity: An Encyclopedia of Advertising and Printing, and All that Pertains to the Public-Seeking Side of Business* (New York: Publicity Publishing, 1897), 655.

14. E. L. Godkin, editor of *The Nation*, for example, complained vehemently of "chromo-civilization" ("Chromo-Civilization," *The Nation*, September 1874, pp. 201–202). For further discussion of chromo reproductions of high art as democratizing and therefore challenging high culture entitlement, see Miles Orvell, *The Real Thing: Imitation and Authenticity in American Culture, 1880–1940* (Chapel Hill: University of North Carolina Press, 1989).

15. *The Literary World*, December 1872, p. 104; quoted in Marzio, *Democratic Art*, 191.

16. The generic chromolithographed image might or might not have any relationship to the advertiser's message. Robert Jay shows two copies of a stock card scene of birds sailing in a wooden shoe, with one overprinted with the name and address of a Fall River, Massachusetts, butcher shop and the other overprinted with the legend "We are bound for the land of promise" from an agent of the St. Paul and Sioux City Railroad in Minnesota. While the latter message creates a tie between the image and the advertised service, the butcher seems not to intend the pictured birds as a sample of his wares and rather has simply put his name and address on an eye-catching card. Jay, *Trade Card*, 37.

17. Trade catalog of A. F. Hunt & Co., "Fashionable Card Printers" of Newburyport, Mass. (n.d.; rare book collection, Winterthur Library). Hunt offers "Assorted packs of Proverbs and Psalms, 25 for 16c, 100 for 40c." The catalog calls the "*Psalm* and *Proverb Cards* . . . splendid cards for Sabbath Schools, and to give away."

18. Laura Ingalls Wilder, *Little Town on the Prairie* (1941; New York: Harper & Row, 1953), 192–193. I wish to thank Nancy Robertson for calling this passage to my attention.

19. Chromolithography was specialized work; because it required the use of large stones brought from Germany, most chromolithograpy was produced in coastal and river cities. Letterpress printing, on the other hand—printing from metal type or engraved cuts—was readily available at local printshops and newspaper offices.

20. W. E. B. Du Bois, *The Souls of Black Folk* in *Three Negro Classics* (New York: Avon, 1965), 214.

21. Twenty years after emancipation, visual representations of African Americans in trade cards positioned black people as intermediaries between commodities and their white purchasers, and as virtual commodities themselves. They were often shown wearing clothing stamped with the name of the product.

22. *The Paper World* 10 (May 1885), 5; cited in Jay, *Trade Card*, 3.

23. Neil Harris, "Color and Media: Some Comparisons and Speculations," in *Cultural Excursions: Marketing Appetites and Cultural Tastes in Modern America* (Chicago: University of Chicago Press, 1990), 320.

24. By contrast, the four-color-process printing in current use (a refinement of a process used since the 1920s) depends on photographically breaking down a preexisting color image into versions of the three primary colors plus black, and then reconstituting the image by printing the four screened color separations on top of one another. The eye integrates the fine dots of which the entire image is composed to read it as a continuous tone, full-color picture. While metallic or fluorescent inks are occasionally run as additional colors, they would not make the image more lifelike.

25. Fowler, *Fowler's Publicity*, 654–655.

26. Chalmers Lowell Pancoast, *Trail Blazers of Advertising: Stories of the Romance and Adventure of the Old-Time Advertising Game* (New York: Frederick H. Hitchcock, 1926), 53.

27. The point is Robert Jay's in *Trade Card*, 91.

28. The term *salvage art* is Elaine Hedges's, from her article "The Nineteenth Century Diarist and Her Quilts," in *American Quilts: A Handmade Legacy*, ed. L. Thomas Frye (Oakland: Oakland Museum, 1981), 58. Hedges notes the similarity of quiltmaking and scrapbook compiling. Writing in a period less admiring of quilts, Janet Ruutz-Rees in 1883 similarly compares scrapbooks and quilts: "The scrap-book proper is like a piece of patchwork, made up of odds and ends." Ruutz-Rees, *Home Occupations*, 98.

29. Lydia Maria Child, *The American Frugal Housewife*, 16th ed. (1835) 3. Quoted in Lynn Turner Oshins, *Quilt Collections: A Directory for the United States and Canada* (Washington, D.C.: Acropolis, 1987).

30. Willa Cather, *My Ántonia* (1918; New York: Signet, 1994), 88–89.

31. See Simon J. Bronner, "Object Lessons: The Work of Ethnological Museums and Collections" in Simon J. Bronner, *Consuming Visions: Accumulation and Display of Goods in America 1880–1920* (New York: Norton, 1989), on the development of museum displays of artifacts. See Jay Mechling, "The Collecting Self and American Youth Movements" in the same volume for discussion of the place collecting held in late-nineteenth century American culture.

32. Ruutz-Rees, *Home Occupations*, 89.

33. Both ibid., 92.

34. Ibid., 89.

35. Arthur Penn, *The Home Library* (New York: D. Appleton, 1883), 80. The name "Arthur Penn" was Brander Matthews's pen name at the satiric monthly *Life* in the 1880s; it is possible that the prolific Matthews is the author of this work as well. (*Life* was a humor magazine until 1936, when Henry Luce bought it to use the title for his new photojournalism monthly.)

36. Steven M. Gelber, "Free Market Metaphor: The Historical Dynamics of Stamp Collecting" in *Comparative Studies in Society and History* 34, no. 4 (October 1992): 743, 747.

37. Henry T. Williams and "Daisy Eyebright," *Household Hints and Recipes* (Boston: Peoples Publishing, 1884), 132.

38. "Editor's Study," *Harper's New Monthly Magazine*, September 1892, p. 639.

39. Arthur Penn, *Home Library*, 80–82.

40. E. W. Gurley, *Scrap-books and How to Make Them: Containing Full Instructions for Making a Complete and Systematic Set of Useful Books.*(New York: Authors' Publishing, 1880), 52.

41. Ibid., 5–6.

42. Among the blank scrapbooks manufactured was the Mark Twain Scrapbook, offered in 1892 as a subscription premium by *Youth's Companion*. Its pages were imprinted with water-soluble glue, allowing material to be moistened and then stuck down, "so that the usual annoyance of pasting the clippings is wholly overcome." The editors make the usual suggestion that selecting material for the scrapbook "is a great help in cultivating the taste. In after years, such a collection will become invaluable." The book was available for one new subscriber plus twenty cents postage and packing, or for eighty cents outright. *Youth's Companion*, October 27, 1892, p. 531. Some compilers, perhaps without this extra cash, made old catalogs or printed booklets into scrapbooks.

43. The covers of the Liebig scrapbooks in the collection of the Metropolitan Museum of Art establish the connection with the advertiser. One, embossed with an elaborate art nouveau pattern of lilies, reads "Chromos de la cie Liebig" (1978.549); another, "Chromos de la Compaigne Liebig" (1978.534). The German album is stamped "Album für Liebig Bilder" (1971.577.1). The compiler of the German album collected cards of other manufacturers' series as well—at least two other advertisers made cards to fit the same slots. All the Liebig's cards were evidently made especially for Liebig's, not as generic cards, but although they carry the name of the product at the bottom, none of the scenarios illustrated involve the product.

44. Pancoast, *Trailblazers,* 53.

45. The arrangements of cards I discuss here are not present in every scrapbook I looked at, nor are they often consistent from page to page within a single scrapbook. While the visual patterns, apparent narratives, and other groupings I've discerned and discuss are necessarily speculative, such arrangements appear in a large enough number of these books to be convincing evidence that these are not haphazard occurrences, but that young people played out their ideas about how cards could or should be arranged.

46. Thelma Mendsen Collection, Winterthur Library, scrapbook 70x1.35. Wells's own unlikely connections between its products include one card, headed "Wells' Health Renewer," in which a landlady complains to her boarder. He sits at the table pouring Health Renewer into a cordial glass, while a boy chases rats:

> *Mrs. Hash*: True! I did agree to board you for $10 a week but I didn't know you were going to take "Wells' Health Renewer" before every meal.
> [*He*:] Well Madame, if you will purchase a box of "Rough on Rats" and clear out all the rats, mice, flies, mosquitoes, roaches and bed bugs, I will pay you three per week extra.

47. Scrapbook 3365 70x1.14 in the Thelma Mendsen Collection, Winterthur Library. Multiple interpretations are possible for such an arrangement. A large card for a Hartford sewing machine ringed by smaller calling cards and reward-of-merit cards could be a grouping of calling cards of owners of sewing machines with the card of the machine, or people the compiler knows in Hartford, or fellow members of a sewing circle, or people who use her family's machine; the reward cards might have been given for good sewing.

48. Maxine Carol Friedman, "Home, Home, Sweet Sweet Home: The Trade Cards of the New Home Sewing Machine Company" (M.A. thesis, University of Delaware, June 1984).

49. Penny McMorris, *Crazy Quilts* (New York: Dutton, 1984), 11.

50. Abba Goold Woolson, *Women in American Society* (Boston: Roberts Brothers, 1873), 235. Quoted in Abelson, *When Ladies*, 36. A story based on this preoccupation is A. Carroll Watson Rankin's "The Quest of a Nile-Green Collar," *St. Nicholas*, August 1905, pp. 889–891.

51. Scrapbook 3365 70x1.6 in the Thelma Mendsen Collection, Winterthur Library.

52. I will be using the generic "she" here for convenience; I have no information on the sex of this compiler.

53. Lina B. Beard and Adelia B. Beard, *The American Girls' Handy Book: How to Amuse Yourself and Others* (1887; New York: Scribner's, 1898), 395.

54. The suggestion also hints that some adults may have been disturbed by children's play with the materials of advertising in ways that left the materials connected to their advertising purpose. Trade card advertisers made relatively little use of nursery rhymes, despite their familiarity.

55. In fact, a quick way to steer children away from playing with ads would have been to give them sheets of die-cut embossed scrap on appropriate topics, seen, for example, in figure 1-5. Some scrapbooks *are* made entirely from such items and are far more decorous and constrained than some of the advertising trade card scrapbooks.

56. Ellen Seiter, *Sold Separately: Children and Parents in Consumer Culture* (New Brunswick, N.J.: Rutgers University Press, 1993), 11.

57. Ruutz-Rees, *Home Occupations*, 99.

58. The compiler of the Strong Museum's scrapbook 86.9463 placed Pinkham's card with other portrait cards of similar format—including one of a maker of flavoring extracts and one of President Garfield.

59. Scrapbook 3365 70x1.4 in the Thelma Mendsen Collection, Winterthur Library. Most of the cards from retailers are from Staten Island and New York, suggesting that she lived on Staten Island. Since the book was vandalized at some point, the "Anna Skinner" card may not mean that she was the compiler.

60. In a typical example of trade cards' casual racism, on the Wheeler's card, one stereotyped Chinese figure says to another, "That's what Melican [American] women use to make fat babies!"

61. Scrapbook 3365 70x1.4 in the Thelma Mendsen Collection, Winterthur Library.

62. "Flowers and Their Sentiments," in Richard A. Wells, *Manners of Culture and Dress of the Best American Society* (Springfield, Mass: King, Richardson, 1891). Cards that appear to be nonadvertising cards in other books specifically refer to language of flower lore: one, for example, with forget-me-nots and a fully open yellow rose carries the message "Love has a language hid from view, / In these fair Flowers it speaks to you." On the page following is a greeting card that calls attention to the connection between the flower's name and its meaning. Embellished with a sprig of forget-me-nots, it is imprinted, "How could I forget thee, / If thou forget me not."

63. Writers on the earlier, clipping-type scrapbooks called their selection and preservation of materials personal and revealing. Ruutz-Rees, *Home Occupations*, believed scrapbooks were "true histories of our inmost selves" (98), as we saw in the epigraph to this chapter, while Gurley, *Scrap-books*, believed that her scrapbooks, made up largely of newspaper clippings, were "such true interpretations of my own feelings, that they show the secret history and aspirations of my soul" (17).

64. L. G. Quackenbush, "The Ladies—A Philatelic Toast," *Pennsylvania Philatelist* 6 (June 1894): 331. Quoted in Gelber, *Free-Market Metaphor*, 749.

65. For a full discussion of scrapbooks made for missionary work, see my "Commercial Fiction: Advertising and Fiction in American Magazines, 1880s to 1910s" (Ph.D. diss. University of Pennsylvania, 1992).

66. Among the premiums given by the Larkin company, makers of soap and other household goods, were albums with a letter inside suggesting that the recipient praise the company's products to friends. Mary E. Moore, "Advertising Cards of the 80s in Upstate New York," *New York History* 30 (October 1949): 458.

67. Edward Bok, *A Man from Maine* (New York: Scribner's, 1923), 17.

68. Jackson Lears discusses some of the complex imagery of trade cards particularly as representations of abundance. See his *Fables of Abundance: A Cultural History of Advertising in America* (New York: Basic, 1994), esp. 102–113.

69. Friedman, "Home, Home," 28–29.

70. In fact, they often make use of the nonverbal scenes that Roland Marchand has identified as social tableaus, which didn't dominate magazine ads until the 1920s. In

contrast to trade cards, most magazine advertisers of the early 1890s relied primarily on verbal appeals and considered illustrations secondary and dispensable.

71. Scrapbook 70x1.9, 34 in the Thelma Mendsen Collection, Winterthur Library.

72. Rosalind Williams, *Dream Worlds: Mass Consumption in Late Nineteenth-Century France* (Berkeley: University of California Press, 1982), 69. For more on the esthetics of excess in Victorian America, see Orvell, *Real Thing,* 42–44

73. Shopping was also integrated with child care: women brought their children shopping with them, and savvy storeowners made sure the store appealed to the child as well, knowing that mothers were more likely to go where the children would be more contented. Girls and young women presumably also went with their mothers to the stores as apprentice shoppers of sorts, although interestingly this role seems not to have been noticed by the advertising departments of women's magazines, which, even in the 1910s, imagined women newly minted as shoppers at marriage.

74. Scrapbook 86.9461 in the collection of the Strong Museum; information on Blackburn and her family is from Strong Museum records. Dates on cards show that she continued keeping the scrapbook until she was at least nineteen.

75. The department stores were also called bazaars; they were part of the same stream of nineteenth-century celebration of the exotic as expositions and worlds fairs, in which the world was miniaturized and turned shopping plaza. Trade card advertisers deliberately appropriated the use of the international as well, issuing such series as Singer Sewing Machine's pictures of people of many countries in national dress sewing with the machines, or Clark's Thread company's series of cards showing postage stamps of different countries, in which the end label of a thread spool figures as another stamp. A stock card series used by both the Great Atlantic and Pacific Tea company and James Kirk and company's Chicago Soaps also showed children in national dress holding a stamp from their country—presumably to enforce the idea of international scope. Both series are in scrapbooks in Album 12, the Waldron collection, Winterthur Library.

76. Girls' play-replication of shopping also took the form of toy shops, of increasing interest later in the 1800s: as a history of toys notes, "the commercially-produced doll's shop as we think of it today is mainly post-1850" (38). The toy manufacturer's attention to detail meant duplicating the packages as well as the goods: "One American grocer's shop, described as the 'Choice Family Grocer's,' contains American-made boxes of labeled half-gross David's Prize Soap—'No Mistake' and 'Rising Sun Stove Polish.'" (The latter, if not the others, is an actual brand name.) Constance Eileen King, *The Encyclopedia of Toys* (New York: Crown, 1978), 41.

77. "Editor's Study," *Harper's,* September 1892, p. 639.

78. Scrapbook 70x1.47 in the Thelma Mendsen Collection, Winterthur Library. Although other articles intended for mainly male use appear on trade cards in the scrapbooks, the cigarette and tobacco cards that came inside the package are relatively rare in scrapbooks, presumably because Papa did not pass such cards along.

79. Williams, *Dream Worlds,* 67.

80. While this process of visual seduction applies most dramatically to department stores—which often had their own trade cards—it applies as well to the grocery and to the other retailers who distributed the cards for nationally available merchandise. While grocery stores at this point were not self-service, the change to branded merchandise meant that goods were displayed in quantity in their packages along the store shelves, with the labels acting to screen the goods themselves with a fantasy of the goods.

81. Gurley, *Scrap-books,* 56.

82. For a development of this insight, see Kimberly Rich, "A Victorian Woman's Material World: The Life and Legacy of Mary Cowgill Corbit Warner" (M.A. thesis, University of Delaware, 1989).

83. Susan Stewart classifies the scrapbook as a souvenir because its parts refer "metonymically to a context of origin or acquisition." In the collection, on the other hand, "each item is representative and works in combination toward the creation of a new whole." As the material of my chapter makes clear, the trade card scrapbook straddles both these definitions. Stewart, *On Longing: Narratives of the Miniature, the Gigantic, the Souvenir, the Collection* (Baltimore: Johns Hopkins University Press, 1984), 152.

84. Williams, *Dream Worlds*, 71.

85. We will see this develop further in the magazines, where, for example, an ad for a long train trip may have run every month, though a consumer might participate only once a year by going on such a trip. Nonetheless, the repeated ad defined a community of potential trip takers, among whom the reader was continually asked to include or exclude him or herself.

86. Ruutz-Rees, *Home Occupations*, 92.

87. Rachel Bowlby, *Just Looking: Consumer Culture in Dreiser, Gissing and Zola* (New York: Methuen, 1985), 32.

Chapter 2. Training the Reader's Attention: Advertising Contests

1. The St. Nicholas League, "'Advertising-Patchwork' Competition. Report of Judges," *St. Nicholas*, August 1902, ad p. 11.

2. The Lord & Thomas agency in Chicago started a "record of results" department as early as 1900 to track sales results from ad placements in different newspapers and magazines. Stephen Fox, *The Mirror Makers: A History of American Advertising and Its Creators* (New York: William Morrow, 1984), 60–61.

3. *Profitable Advertising*, February 1904, p. 975; emphasis added.

4. Similarly, a 1903 venture into market research by J. Walter Thompson's agency, in soliciting reader reactions to ads, seems to have focused on how people "liked" ads. Study cited in Jackson Lears, *Fables of Abundance: A Cultural History of Advertising in America* (New York: Basic, 1994), 211.

5. Janet E. Ruutz-Rees, *Home Occupations* (New York: D. Appleton, 1883), 92.

6. For a summary of currents in advertising psychology into the 1920s, see A. Michal McMahon, "An American Courtship: Psychologists and Advertising Theory in the Progressive Era," *American Studies* 13, no. 2 (fall 1972): 5–18. Rachel Bowlby notes that psychologists in the early twentieth century were "laying out areas of investigation which overlap to a striking extent with those that concerned advertisers from a pragmatic point of view." Bowlby, *Shopping with Freud* (London: Routledge, 1993), 96.

7. Walter Dill Scott sets out various schemes for establishing and strengthening the reader's memory of an advertisement in his *The Psychology of Advertising: A Simple Exposition of the Principles of Psychology in Their Relation to Successful Advertising* (1908; Boston: Small, Maynard, 1917). Much of the material in the book appeared in earlier articles and lectures from 1900 on.

8. *A Good Line of Advertising*, St. Nicholas League Advertising Booklet no. 5, p. 2; bound into *Profitable Advertising* 13, no. 8 (January 1904): following 817.

9. Raymond Williams points out that the word *character* originated with the Greek word for an engraving or impressing instrument, which led to the modern usage that denotes a letter of the alphabet. Up through the sixteenth century, the word *char-*

acter was used to represent an impressed sign or symbol, and later it came to describe an essential nature or quality. *Character* is therefore both fixed and readable. Warren I. Susman notes that in the nineteenth-century United States, a vision of self embodied in the term *character* was to lead, in its ideal version, to self-control and self-mastery. This vision of self was eventually displaced by *personality*, a version of the self more congenial to advertisers with its emphasis on self-fulfillment, self-expression, self-gratification, and being well-liked. Williams, *Keywords: A Vocabulary of Culture and Society*, revised ed. (New York: Oxford, 1983), 234, and Susman, *Culture as History: The Transformation of American Society in the Twentieth Century* (New York: Pantheon, 1984), 280.

10. Walter Dill Scott, *The Psychology of Advertising in Theory and Practice* (1902–1903; Boston: Small, Maynard, 1921), 218.

11. Ibid., pp. 92–93.

12. "The Proceedings of the Sphinx Club," *Printer's Ink,* March 2, 1904, p. 59.

13. MacGregor Jenkins, "Human Nature and Advertising," *The Atlantic Monthly*, 1904, p. 397.

14. *Selling Forces* (Philadelphia: Curtis, 1913), 161–162. This book was addressed to potential advertisers in *Ladies' Home Journal* and *Saturday Evening Post*.

15. *Good Line of Advertising*, 3.

16. Ibid., 4.

17. *St. Nicholas*, January 1905, ad p. 16.

18. *Good Line of Advertising,* 7–8.

19. "$10 Prize Offer," *Woman's World and Jenness Miller Monthly*, January 1897, p. 25.

20. *Profitable Advertising*, May 1905, ad insert following p. 1328.

21. E. B. White, "The St. Nicholas League" in the column "Onward and Upward with the Arts," *The New Yorker* December 8, 1934, p. 42.

22. *St. Nicholas* data from "Alice in Blunderland: To St. Nicholas Advertisers of the Past, Present, Future," *Profitable Advertising*, December 1902, p. 12; *Youth's Companion* data from Frank Luther Mott, *A History of American Magazines* (1938; Cambridge: Harvard University Press, 1957), 2:268. Moreover, to another writer on a farm in rural Pennsylvania in the 1880s, even *Youth's Companion* seemed remote and genteel next to the far more popular dime novels to which he was accustomed. Mark Sullivan, *The Education of an American* (New York: Doubleday, Doran, 1928), 72.

23. *Good Line of Advertising*, 12.

24. "The St. Nicholas Competition," *Profitable Advertising*, March 1903, p. 873. The unsigned article's close echoing of *St. Nicholas's* own publicity suggests that it may have been dictated by the magazine, a frequent advertiser in *Profitable Advertising*.

25. "Instructions for competition 108," *St. Nicholas*, December 1910, ad p. 24.

26. *Profitable Advertising*, February 1905, ad following p. 968. This claim appeared even before the rules were changed to allow collaboration and adult entries.

27. "A Century of Questions," *St. Nicholas*, December 1904, ad p. 22.

28. "Report on Competition 43," *St. Nicholas*, April 1905, ad p. 14.

29. Margaret Johnson, "The Corner Cupboard," *St. Nicholas*, July 1905, pp. 777–784.

30. Despite this, nearly every issue carried news that a child's prize had been revoked because it was found to have been given for copied work.

31. *St. Nicholas*, April 1911, ad p. 6. This competition page announces that Don M. Parker has been installed as advertising manager; from March on the style of writing changes to this punchy, breezy approach.

32. "Vacation Advertising Competition of the St. Nicholas League," *St. Nicholas*, June 1902, ad p. 10.

33. "Report on Competition 40," *St. Nicholas*, December 1904, ad pp. 26–28.

34. *St. Nicholas*, April 1911, ad p. 6.

35. Because advertising was confined to separately paginated sections in the front and back of most magazines until the 1910s, most libraries stripped out the advertising before binding the magazines for permanent use. Bound volumes available through *St. Nicholas* in distinctive covers also had the ads stripped out.

36. Other magazines encouraged readers to help the magazine out as sellers and promoters: *Ladies' Home Journal* and *Youth's Companion* were among the magazines that offered premiums and prizes for signing up subscribers.

37. *St. Nicholas*, September 1911, ad p. 8.

38. Scott, *Psychology of Advertising*, 14.

39. *St. Nicholas*, March 1911, ad p. 6.

40. *St. Nicholas*, July 1911, ad p. 12.

41. Ibid.

42. Nelle M. Mustain, *Popular Amusements for in and out of Doors, Embracing Nine Books in One Volume, The Whole Comprising a Charming Collection of Games, Sports for Health and Beauty, Instructive Amusements and Miscellaneous Helps for Both Young and Old, for the Home, the Church, and the School* (No city given, Lyman A. Martin, 1902), 15–16. A 1905 book calls a closely related memory game "Advertisement Items." Mrs. Herbert B. Linscott, *Bright Ideas for Entertaining* (Philadelphia: George W. Jacobs, 1905), 4.

43. *St. Nicholas*, April 1901, ad p. 13.

44. "Alice in Blunderland," 5.

45. Linscott, *Bright Ideas*, 4.

46. M. T., "An Advertising Entertainment," *Mahin's Magazine*, January 1903, pp. 47–48.

47. Ibid.

48. "Report on Competition 40," *St. Nicholas*, December 1904, pp. 26–28. Entrants were to have one of the members of the family "be unique and . . . advertise an especial article, as the 'Belle Chocolatier' advertises Baker's Chocolate."

49. Similarly, an ad contest in *The Mother's Magazine* in 1909 asked women to read through a magazine's ads, find four listed phrases buried in them, and identify the ads in which they appeared. But finding the phrases was a passport to another, presumably more engaging, portion of the task: entrants were to write an anecdote on "which of your prejudices creates the most amusement among your friends?" ("October Prize Contest," *The Mother's Magazine*, October 1909, p. 74). By 1909, the advertising portion of the contest seems perfunctory, as though it were no longer thought to supply entertainment value in itself, and yet evidently readers were so accustomed to this combination of requirements that no justification for the advertising portion is given. It is presented as natural that the reader be required to look through ads before going on to write, perhaps just as she looks through the ads before turning to a story or article.

50. "Prize Ad. Stories," *Ladies' World*, March 1904, p. 35.

51. Nina Auerbach, "Victorian Womanhood and Literary Character" in *Woman and the Demon: The Life of a Victorian Myth* (Cambridge: Harvard University Press, 1982), 212. Auerbach characterizes Victorian literary criticism in general along these lines as "a reverent scrutiny of a grand freestanding character extracted from its surrounding textual medium" (190). See also the reproductions of "Dickens's Dream" and "The Empty Chair" on pp. 202 and 204. Her discussion of the separability of character from the work emphasizes the particularly Victorian interest in seeing literary figures as transcending the work, owning an immortality that gives them life beyond the author or reader. When crossed with the already deracinated advertising figures, this

tradition of separating characters from the works in which they appear emerges as a precursor of such late twentieth century character-licensed figures as Strawberry Short-cake or My Little Pony: characters which are first created, then sold as toys or on greet-ing cards, and finally placed in stories and backgrounds to generate sales and markets in additional media.

52. Unsigned, "From 'P.A.'s' Point of View: Advertising Phrases and Characters," *Profitable Advertising*, December 1905, p. 712.

53. Both quotes from ibid. *Mahin's*, another advertising trade journal, noted other cartoons adapted from the Gold Dust Twins theme. Unsigned, "Strong Advertising Reflected by the Cartoonists," *Mahin's*, April 1902, p. 20. The frequent use of black fig-ures as representatives of products began with trade cards, if not earlier, where they very often wore clothes or used tools that were marked with the name of the product. In many of the trade card scenarios in which they appear, the black figures speak with spe-cial authority about commodities or are shown (sometimes comically) making choices about seemingly unrelated realms of their lives based on commodities. A trade card for Standard Shoe and Boot Screws, for example, shows a black woman throwing over a beau who does not use the product for another who does; his shoes are emblazoned with the product name. The notion of choosing a mate on the basis of his use or nonuse of a product, though later developed into classic advertising scenarios meant to be taken seriously, was still a fanciful and comic one at this point; the trade card floats it as a pos-sibility but projects it onto absurdly dressed minstrel-show blacks.

Beyond the intended comic element, it is evident that less than twenty years after African Americans stopped being commodities themselves, figures of blacks were used in advertising to speak from the world of commodities. Characters such as the Gold Dust Twins, appearing in print ads in the 1890s, spoke not only *from* the world of com-modities but also *as* the commodity themselves. The implicit promise of the Gold Dust Twins ads was that the product would supply the buyer with the equivalent of the labor of two tireless, industrious children: slaves in a box. Not surprisingly these ads coin-cided with a fashion for magazine stories romanticizing the antebellum South, as will be discussed further in chapter 3.

54. "Prize Ad. Stories," *Ladies' World*, March 1904, p. 35.

55. Ibid.

56. Ibid.

57. The reference to eight hundred entries appears in *Ladies' World*, May 1904, p. 35. An article on the contest in *Judicious Advertising and Advertising Experience* (July 1904, pp. 66–67), however, interpreted this to mean that "a careful inspection of the tally shows that all the ads in the paper were thoroughly read by thousands of people"—meaning perhaps that there were many collaborative entries, and pos-sibly also assuming that many people would have begun but not completed an entry. The magazine's total circulation at this time is given as "500,416 per issue" on a January 1904 list in *Profitable Advertising* of "papers that are willing to have their circulations investigated."

58. Beulah Putnam, "The Adventures of the 'Healthy Baby' and the 'Happy Young One,'" in "'Healthy' and 'Happy'—An Interesting Advertising Contest and Some Facts and Figures Concerning It—One of the Winning Stories in Verse," *Profitable Advertising*, May 1904, pp. 1290–1291.

59. Album of Mary Eliza Bachman, entitled "The Friendly Repository and Keep-sake." In the Joseph Downs Collection, Winterthur Library.

60. In "Bible Salad," which appears in several books of amusements, players iden-tify biblical verses written on slips. The leader finally reads the correct list and the player

with the fewest mistakes is the victor. Amos R. Wells, *Social Evenings: A Collection of Pleasant Entertainments for Christian Endeavor Societies and the Home Circle* (Boston and Chicago: United Society of Christian Endeavor, 1894), 24–25.

61. Roland Marchand, *Advertising the American Dream: Making Way for Modernity, 1920–1940* (Berkeley: University of California Press, 1985).

62. Historians of the book have traditionally argued that a shift in reading practices occurred in the west with the burgeoning availability of cheap printed matter: instead of reading the Bible or a few religious texts repeatedly and intensively, and seeing them as containing all meaning, readers learned to read extensively, taking in large amounts of printed information, studying it less closely, and learned to skim and to skip along the way. More recent writing in this area suggests that there is no single point at which this shift occurred and that both practices could coexist. See, for example, Carlo Ginzburg, *The Cheese and the Worms: The Cosmos of a Sixteenth-Century Miller* (Baltimore: Johns Hopkins University Press, 1980). Advertisers often addressed the problem of their ads getting lost in the sea of reading matter by taking up what was called "extensive advertising," hoping that if the ad bobbed up often enough it would be remembered.

63. Marion Sackett, *The "Heart Beats" of a Great City* (Coal Oil Johnny's Petroleum Soap pamphlet). No date, but the story is set in 1894 or 1895 and clothing in the illustrations seems consistent with that. In the Bella Landauer collection, scrapbook 8C; New-York Historical Society.

64. Ibid.

65. Ibid.

66. *St. Nicholas*, March 1911, ad p. 6.

67. "Report on advertising competition 47," *St. Nicholas*, September 1905, ad p. 10.

68. "Competition 44," *St. Nicholas*, March 1905, ad p. 10.

69. White, "St. Nicholas League," 42. Among the later-famous who won nonadvertising St. Nicholas League competitions as children, White found Vita Sackville-West (writing about Knole), Ring Lardner, Conrad Aiken (four times), Robert Benchley (for a drawing), Edna St. Vincent Millay (twenty times), Elinor Wylie, Cornelia Otis Skinner, Janet Flanner, Alan Seeger, William Faulkner, Norman Geddes (for a drawing), and many others. The only later famous winner of an advertising contest I've recognized is Robert Moses.

70. Lucy Maude Montgomery, *Anne of the Island* (1915; New York: Bantam, 1976), 88.

71. Ibid.

72. Ibid.

73. Putnam, "Adventures of the 'Healthy Baby.'"

Chapter 3. "The Commercial Spirit Ḩas Entered Jn": Speech, Fiction, and Advertising

1. Edwin L. Sabin, "A Rhapsody in Realism," *Profitable Advertising*, September 1903, p. 315.

2. Ellis Parker Butler, "The Adventure of the Princess of Pilliwink," *Judicious Advertising and Advertising Experience*, August 1904. Reprinted in *Perkins of Portland: Perkins the Great* (Boston: Herbert B. Turner, 1906), 116–135. Stories in this volume originally appeared in the *Saturday Evening Post* and, along with other uncollected Perkins stories, in *Judicious Advertising and Advertising Experience*.

3. Samuel Hopkins Adams, "The New World of Trade" *Collier's*, 22 May 1909, p. 13.

4. E. C. Patterson, "Advertising Bulletin Number 4: The Cost of Advertising," *Collier's*, 22 May 1909, p. 5; emphasis in the original.

5. A smaller countertrend followed the practice of patent medicines, which, to attract buyers, claimed to treat a long list of diseases and asserted that the product would serve multiple uses—that Pearline laundry soap, for example, was fine for bathing baby as well.

6. Ad for Pears' soap, *Cosmopolitan*, probably January 1892, ad p. 1.

7. These are social classifications as well: the consumer not only recognizes the soap suited to a particular task but also recognizes in what class of home the soap is likely to be found.

8. Judith Williamson, "Three Kinds of Dirt," *Consuming Passions: The Dynamics of Popular Culture* (London and New York: Marion Boyars, 1986), 225.

9. Although common law had protected trademarks before this, the first U.S. trademark law was not passed until 1881, setting off decades of trademark suits.

10. Herbert W. Hess, *Productive Advertising* (Philadelphia: Lippincott, 1915), 164.

11. Manufacturers as a group use a method to protect trademarks that paradoxically highlights the "naturalness" that relentless promotion has achieved for commercial neologisms and demonstrates how easily they replace generic words. The trade organization, the United States Trademark Association sends out an authoritative-sounding editorial style sheet to publishers and editors in an effort to establish a paper trail of protection for brand names. The sheet instructs editors to "Capitalize trademarks and use them as adjectives with the generic term of the product. For example, Kleenex facial tissues . . . Instamatic camera. . . . Do not use the trademark without the generic term, and always capitalize the trademark." Other recommended terms are "Chiclets chewing gum . . . Fig Newtons cakes . . . Jell-O gelatine dessert . . . Metrecal brand dietary for weight control . . . Ouija talking board sets . . . Saran Wrap brand of plastic film . . . Teflon fluorocarbon resins." *Trademark Stylesheet* No. 1 A (New York: U.S. Trademark Association, no date, in circulation in 1991); the association has since been renamed the International Trademark Association. In other words, while companies wish their brand names to move into *speech* as household words, to protect their interests in the words, they must more rigidly police their use in writing, or at least announce the rules, to keep from losing them. They therefore reinsert the generic into the phrase at this point.

For discussion of some of the further cultural ramifications of trademark law, see Jane M. Gaines, *Contested Culture: The Image, the Voice, and the Law* (Chapel Hill: University of North Carolina Press, 1991).

12. *St. Nicholas*, April 1901, ad p. 13.

13. Richard Bushman explores the architectural exposition of this division in middle-class houses in nineteenth-century America, where the "crucial characteristic of a refined house was a parlor, free of work paraphernalia and beds. . . . Only genteel occupations like sewing, reading, or visiting went on in the parlor and sitting rooms. . . . Business was not to show itself in the genteel portions of the house." Bushman, *The Refinement of America: Persons, Houses, Cities* (New York: Knopf, 1992), 251, 262.

14. Advertisement, *System: The Magazine of Business*, April 1906, unpaginated.

15. "The St. Nicholas Competition," *Profitable Advertising*, March 1903, p. 873.

16. "At the Coaching Parade," *Art in Advertising*, April 1890, p. 39. The early 1890s issues of this journal in general reflect little of the single-minded seriousness about business found in the other advertising trade journals; self-mockery is more prominent.

17. Ibid., p. 37.

18. Diana and Geoffrey Hindley, *Advertising in Victorian England* (London: Wayland Publishers, 1972), 46.

19. This ad appeared often, including in *The Atlantic Monthly*, November 1893, ad p. 38. The slogan accompanied several other ads as well, including a beach scene of a white nymph emerging from a sea shell, startling a small black boy with the greeting. The racist overtones of this ad are better understood when placed in context: black characters appeared in toilet soap ads at that time only as a racist joke, to demonstrate the ads' claim that black skin could be scrubbed white by the product. A Pears' ad using the motif seems to have appeared primarily in England. But in this Pears' ad, clearly the white nymph *has* used Pears' and the little boy hasn't: he's still black.

20. Frank Presbrey calls Pears' one of the first four companies to undertake systematic, widespread advertising. The others were Royal Baking Powder, Sapolio, and Ivory—and these are notably companies whose names and slogans appear frequently in the crossovers to nonadvertising space discussed here. Presbrey, *The History and Development of Advertising* (Garden City, N.Y.: Doubleday, Doran, 1929), 339.

21. Amelie Rives (later Amelie Rives Chanler Troubetzkoy), *The Quick or the Dead?* (Philadelphia: Lippincott, 1888), 228–229.

22. *Lippincott's* practice of printing entire novels at once was successful for at least some years, and "its success has induced other monthlies to adopt it," according to an 1891 article, which doesn't say which magazines did it. Junius Henri Browne, "American Magazines," *Author*, January 1891. The Rives book was reviewed on the basis of its appearance in the magazine.

23. *The National Cyclopaedia of American Biography* (New York: James T. White, 1927).

24. Amelie Rives, *Quick or the Dead?*, 222.

25. Ibid., 97.

26. A similar distinction operated in this period between the acceptable, quasi-natural category of nonvisible materials such as lotions and skin creams, which could be seen as simply part of skin care, and visible makeup. Skin care products were also an element in a process of transforming the *interior* self through "beauty culture" relaxation regimens, thereby transforming appearance, in line with an ideology that asserted that beauty was an outward manifestation of inner value. See Kathy Peiss, "Making Faces: The Cosmetic Industry and the Cultural Construction of Gender, 1890–1930," *Genders* (spring 1990): 143–169.

27. Henry Ward Beecher's affiliation with the soap is reinforced on a large trade card, in an 1886 series featuring portraits of each of the endorsers. The series carries endorsements from Lillie Langtry, Adelina Patti, Mary Anderson, and Sir Erasmus Wilson, Late President of the Royal College of Surgeons, as well. Bella Landauer Collection, New-York Historical Society. Box: Personal Grooming Aids, 3; Toilet soaps.

28. Rives, *Quick or the Dead?*, 178.

29. Christopher P. Wilson classifies the audience for magazines like *Lippincott's* as "mostly Northeastern, mostly well-to-do" in his *The Labor of Words: Literary Professionalism in the Progressive Era* (Athens: University of Georgia Press, 1985), 43.

30. Frank Munsey, "Derringforth," *Munsey's* 1893–1894. Other stories include "My Mare Sally," by Phillips McClure (*Munsey's*, October 1893, pp. 12–15), in which the visiting northerner fails to get the southern woman. Paul Buck has pointed out that in the 1880s what he calls "the reconciliation motif" was "conventionalized into a plot which married a Northern hero to a Southern heroine"; the union resulted in "Americanism." Buck, *The Road to Reunion: 1865–1900* (New York: Vintage, 1959), 223. Henry James's *The Bostonians*, in which the southern man visiting the North carries back a northern woman, is a possibly self-conscious reversal of this trope.

31. Rives, *Quick or the Dead?*, 229.

32. The racist representation of black figures on trade cards in the recently reunited United States of the 1880s also became a bridge between the white North and South. As brand names became part of national culture, reading blacks out of the consumer role and into the role of commodity had special implications for North-South relations. The recommodified black bodies of the African-American figures on trade cards, who typically wore clothing stamped with the name of the product, for example, or stood in as the trademark for the commodity itself, as in the case of the Gold Dust Twins discussed in chapter 2, served not only as intermediaries between product and consumer, human projections of the product but also as a locus of new agreement between North and South: the commodity status of blacks was reinscribed nationwide through advertising.

33. Nym Crinkle, "The Priestess of Passion: Nym Crinkle discusses Amelie Rives, the Romantic Virginian," *New York World*, 15 July 1888, p. 13. Crinkle was the pen name of book reviewer Andrew Carpenter Wheeler.

34. Unsigned, "A Girl's Experiments: Miss Amelie Rives's Work," *New York Daily Tribune,* 13 May 1888.

35. *Munsey's*, December 1893, p. 319.

36. *New York World*, 23 September 1888, p. 10.

37. The 1901 *Sears and Roebuck Catalog*, for example, "offered electric belts designed to restore men's sexual powers or to help either sex restore lost energies," David Nye notes in *Electrifying America: The Social Meanings of a New Technology, 1840–1940* (Cambridge: MIT Press, 1990), 153.

38. All preceding quotes this paragraph from "Literary Chat," *Munsey's*, December 1893, p. 330.

39. Nathaniel Fowler, *Fowler's Publicity: An Encyclopedia of Advertising and Printing, and All That Pertains to the Public-Seeing Side of Business* (New York: Publicity Publishers, 1897), 455–457.

40. This type of scheme is reported in *The Nation* in 1869. Small specialized trade journals sold both advertising and editorial notice at a joint fee for "a whole column of advertising and half a column of editorial notice, the editorial notice to be furnished by the advertiser." The advertiser could also pay to exclude contradictory information. The reading notice in such papers was not the final object: "It is indeed not published to be *read*; it is only published to be *quoted*." The agent would next buy advertising in the "respectable but struggling dailies or weeklies," stipulating that the paper occasionally "copy and insert semi-editorially extracts from other papers in relation to the new company, which will be kindly furnished by the agent himself. In this way many a newspaper editor, striving to be honest and fair, is yet induced to reprint an outrageous puff from the *Weekly Steamship*, and soothes his conscience with the excuse that every reader sees that the paragraph is an extract from another paper." But the notice in the "respectable" daily or weekly would then be parlayed into mention in the financial pages, and finally editorial endorsement, by "the principal journals in the large cities all over the country, the great leaders of public opinion." Through this complicated but manageable system, then, straightforward advertising could be transformed into editorial notice. "Financial Advertisements," *The Nation* 9, no. 218 (September 2, 1869), pp. 186–187.

41. Unsigned, "Random Notes," *Art in Advertising*, October 1890, p. 137. The reference to Cantaloupe's food is a pun on Mellin's food, which was widely advertised as suitable for infants and invalids in a long-running series of ads featuring pictures of healthy infants, captioned "We are advertised by our loving friends."

42. "Current Notes," *Lippincott's*, March 1888.

43. Unsigned, "May's Triumph," *Ladies' Home Journal*, October 1894, p. 27.

44. Unsigned editorial, "Ethics and Esthetics of Advertising," *Independent,* January 2, 1902, p. 53.

45. Tom Masson, "Patent Advertising Story, Style A—For Mention in This Story, Preferred Position, Etc., Apply to the Hubbub Advertising Company, New York, Chicago, and Boston," *Profitable Advertising,* June 1904, pp. 30–31.

46. Earnest Elmo Calkins, "Fact or Fiction for Advertising," *Judicious Advertising and Advertising Experience,* January 1904, p. 13. Thanks to Roland Marchand for calling this article to my attention.

47. Wilder Grahame, "Serving Ads. with Sauce," *Profitable Advertising and Art in Advertising,* February 1900, pp. 621–622.

48. A number of magazines did in fact run ads—often advertising baking powder—that looked much like articles, in the pages directly after the reading matter. *Art in Advertising* mentions several "four-page reading articles" placed in major magazines, one for Walter Baker's cocoa and one for Hawk-eye cameras, the latter taken in "the three big magazines" (*Scribner's, Harper's, The Century*) in May 1890, costing a total of $3,550.00 for one insertion.

49. "Advertisements in Novels," *Profitable Advertising and Art in Advertising,* October 1899, p. 336. Caine was reported to have looked out for his own interests in the matter by investing in Manx real estate himself.

50. Earnest Elmo Calkins, "Practical Observations on the Practise of Advertising, with Conclusions Drawn from Actual Experience," *Profitable Advertising,* April 1904, p. 1144. The book referred to is Charles Norris Williamson and Alice Muriel Williamson, *The Lightning Conductor: The Strange Adventures of a Motor-Car* (New York: Holt, 1903).

51. I have found no Pears' ads that make either verbal or iconographic reference to the novel. A. A. Cole, consultant archivist at Unilever in London, which now owns Pears', agrees with the *Munsey's* columnist in seeing the Rives notice as only an inadvertent advertisement, not noted or exploited by the company. Cole reports that "a careful search through the Pears' archives, including a volume of press cuttings covering this period," reveals no references to the novel. Personal communication, June 20, 1990.

52. Calkins, "Practical Observations," 1144.

53. William Dean Howells, *A Hazard of New Fortunes* (1890; New York: New American Library, 1965), 281, 190.

54. Butler, "Mr. Perkins of Portland," in *Perkins of Portland,* 12.

55. Advertisement, inside front cover, *Art in Advertising,* June 1890. Even before the ad asserted Fulkerson's verisimilitude, the journal, by running a monthly column signed "Fulkerson," had suggested that he was a reasonable match for a real person. The ad went on to call Fulkerson a character "of which the prototype exists in almost every large publishing house in the city. He comes in contact daily with the men who advertise, and knows all the outs and ins of the business."

56. Butler, "The Adventure of the Lame and the Halt," in *Perkins of Portland,* 44–45. The slogan itself is a take-off on a slogan for Jones Scales: "Jones, he pays the freight."

57. Vaudeville, which shared the magazines' need to keep up with current catchphrases, did incorporate ad slogans. For example, a British theater song based on the Pears' slogan "He won't be happy till he gets it" was used in the 1888 production of *Faust up to Date.* See Ellen Garvey, "Commercial Fiction: Advertising and Fiction in American Magazines, 1880s to 1910s" (Ph.D. diss., University of Pennsylvania, 1992), 243–245.

58. This and preceding quotes from Butler, "Adventure of the Princess of Pilliwink," 126. Butler's satire points up the arbitrariness and absurdity of the slogans themselves, with, for example, his switch from "strength of Gibralter" to "strength of Port Arthur."

59. A satiric story in the trade journal *Art and Advertising* similarly sprinkles ad references throughout for comic effect, though here the slogans and brand names are kept intact, as in the sample story for the *Ladies' World* contest discussed in chapter 2. Other names, however, are changed: Woodyard instead of Rudyard Kipling, perhaps to indicate that we *are* in a fictional universe. The story is illustrated with excerpts from familiar ads as well. A sample: "Amaryllis . . . crossed the almost darkened room, pulled up the James G. Wilson's Venetian blinds, handsomely trimmed with linen, silk or oxidized silver ladders, and looked forth. . . . Her gentle bosom rose and fell with swiftly increasing palpitation (for which Dr. Pierce's Favorite Prescription might well be commended) as she saw the familiar form of Algernon coming down the street." Her father objects to Algernon as a suitor because he is an ad agent. The serial unfortunately appears not to have continued past two installments. "Our Own Serial, Written and Illustrated by Ourselves," *Art and Advertising*, March 1891, April 1891.

60. Mark Crispin Miller's documentation of this practice argues that advertisers' desire to associate their products only with upbeat events and storylines has shaped late-twentieth-century Hollywood filmmaking. He points to the more thoroughgoing saturation with product placements of films made by corporations largely owned by the makers of those products. He finds, for example, glowing treatment both blatantly and subtly accorded Coke in films produced by Columbia Pictures during the period when it was 49 percent owned by Coca-Cola, and a concomitant demonization of Pepsi-Cola in those same films. "Hollywood: The Ad," *The Atlantic*, April 1990, 41–68. The 1991 film *Wayne's World* satirized the practice in a scene in which the characters speak against product placement while ostentatiously displaying brand-named goods, thus acknowledging that the audience was fully aware of the practice. But, as in many commercials of the 1980s and 1990s, in which a metacommercial structure invites the audience to relish its superior knowledge of advertisement, this invitation to take a superior or ironic stance toward advertising in reality undercuts neither the ad message nor the practice.

61. Amy Kaplan, *The Social Construction of American Realism* (Chicago: University of Chicago Press, 1988), 9.

62. Sabin, "Rhapsody in Realism."

63. Letter from a reader, "A Woman, New York, April 7, 1888," in *The Critic,* April 14, 1888, p. 181. It is perhaps also relevant that a written work need not go to special lengths to avoid using brand names. The reader will not be jarred by a mention of *cereal* rather than *Kellogg's Corn Flakes*, and, having read of cereal, is free to imagine a specific type or not. It is perhaps this usurpation of the readers' prerogatives that makes problematic the intrusion into fiction of a brand name when it is not the object of satire.

64. Orson Lowell, quoted in Forrest Crissey, "Ethics as Related to Art and Literature: No. 3—The Shattering of Traditions," *Mahin's Magazine*, May 1903, p. 105. Lowell also suggests that there is no harm to artists in having their work used in ads, or even in creating work especially for ads, as long as the "commercial element is not allowed to obtrude itself to the detriment of the artistic beauty of the picture," and provided, too, that the picture be well reproduced.

65. John Guille Millais, *The Life and Letters of Sir John Everett Millais, President of the Royal Academy* (New York: Frederick Stokes, 1899), 2:190.

66. Letter, Marie Corelli to John Everett Millais, December 24, 1895, in ibid., 190–191.

67. "Miss Corelli's Crusade—She opposes the erection of a Carnegie library at Stratford" in "Literary Chat," *Munsey's*, June 1903, p. 446.

68. Henry C. Sheafer, "A Study in Advertising," *The Arena* 29 (1902): 387.

Chapter 4. Reframing the Bicycle:
Magazines and Scorching Women

1. Maria E. Ward, *Bicycling for Ladies: The Common Sense of Bicycling* (New York: Brentanos, 1896), 12–13.

2. H. O. Carrington, "As to the Bicycle," *American Midwife* 2 (1896): 16; quoted in John S. Haller and Robin M. Haller, *The Physician and Sexuality in Victorian America* (Urbana: University of Illinois Press, 1974), 184.

3. See, for example, Elaine Abelson, *When Ladies Go A-Thieving: Middle-Class Shoplifters in the Victorian Department Store* (New York: Oxford University Press, 1989), and Carroll Smith-Rosenberg, "The New Woman as Androgyne" in *Disorderly Conduct: Visions of Gender in Victorian America* (New York: Knopf, 1985).

4. The rare exceptions to this rule were an already suspect class of women: stage performers who used the high-wheeler in an act. Women could ride the cumbersome carriage-like tricycle sometimes known as a fairy tricycle; riders sat on a bench between two large wheels and steered a small wheel in front. Riding the high-wheeler and the tricycle were seen as complementary gendered activities. See Ellen Garvey, "Commercial Fiction: Advertising and Fiction in American Magazines, 1880s to 1910s" (Ph.D. diss., University of Pennsylvania, 1992).

5. The bicycles on which Lancelot and his knights ride to Hank Morgan's rescue in Twain's 1889 *A Connecticut Yankee in King Arthur's Court* are the manly high-wheelers. An 1880s enthusiast in a bicycling magazine similarly establishes riding the high-wheeler as a masculine and chivalrous pursuit: "gallant knights and true are they; / Full hard they ride, oft brook mischance, / and daring feats full oft essay, / For maiden's praise or tender glance." Basil Webb, "A Ballade of This Age," *The Wheelman* 3, no. 1 (October 1883): 100.

6. Max Vander Weyde, "A Turn of the Wheel," *Godey's Lady's Book,* December 1888, p. 480.

7. See for example Emily Lennox's stories "A Love Game," *Godey's Lady's Book*, October 1888, pp. 295–299, and "The Unbidden Guest: An Episode in a Country House," *Godey's Lady's Book*, November 1888, pp. 349–360. The formula is common in *Munsey's* in the 1890s as well. For a discussion of a similar pattern in British women's novels of the mid-nineteenth century, see Sally Mitchell, "Sentiment and Suffering: Women's Recreational Reading in the 1860s," *Victorian Studies* 21, no. 1 (autumn 1977): 29–45.

8. Paper dolls advertising Columbia Bicycles for Women's Use, 1895. Joseph Downs Collection of Manuscripts and Printed Ephemera, Winterthur Library, 78x317.12.

9. Envoy and Fleetwing: advertisement for the Buffalo Cycle Co., *The Century*, February 1896, ad p. 38. Napoleon and Josephine: Jenkins Cycle Co., *Scribner's*, August 1896, ad p. 34.

10. Frank Presbrey, *The History and Development of Advertising* (New York: Doubleday, Doran, 1929), 410.

11. For a fuller discussion of the bicycle and tourism, see Gary Allan Tobin, "The Bicycle Boom of the 1890s: The Development of Private Transportation and the Birth of the Modern Tourist," *Journal of Popular Culture* (spring 1974): 838–849.

12. For a discussion of the treatment of women's athleticism in the satiric press, see Patricia Marks, *Bicycles, Bangs and Bloomers: The New Woman in the Popular Press* (Lexington: University Press of Kentucky, 1990).

13. Martha Banta also discusses the importance of the image of the woman on the

bicycle, asserting that the visual image of the American Girl as bicyclist helped form new conceptions of what women might do and be. Banta, *Imaging American Women: Idea and Ideals in Cultural History* (New York: Columbia University Press, 1987), 88.

14. Frances E. Willard, *How I Learned to Ride the Bicycle: Reflections of an Influential 19th Century Woman* (Revised ed. of *A Wheel within a Wheel*, 1896; Sunnyvale, Cal.: Fair Oaks Publishing, 1991), 43.

15. Ward, *Bicycling for Ladies*, 12–13.

16. Marks, *Bicycles, Bangs and Bloomers*, 174.

17. Marguerite Merington, "Woman and the Bicycle," *Scribner's*, June 1895, p. 703.

18. Willard, *How I Learned*, 32, 74.

19. See, for example, Arthur Bird, M.D., "Is Bicycling Harmful?" *Godey's Magazine*, April 1896, p. 374, and J. West Roosevelt, "A Doctor's View of Bicycling," *Scribner's*, June 1895, p. 708. A notable exception to this practice is Ward's *Bicycling for Ladies*, with photographs by Alice Austen. Ward not only celebrated the autonomy the bicycle afforded women, but augmented her book with thorough instructions on bicycle repair and adjustment. In a chapter titled "Position and Power," Ward discusses the seat solely in terms of the physical mechanics of riding.

20. Willard, *How I Learned*, 58.

21. E. D. Page, "Women and the Bicycle," *Brooklyn Medical Journal* 11 (1897): 84. See also Thomas Lothrop and William Potter, "Women and the Bicycle," *Buffalo Medical Journal* 35 (November 1896): 348–349.

22. Karin L. F. Calvert, *Children in the House: The Material Culture of Early Childhood, 1600–1900* (Boston: Northeastern University Press, 1992), 114.

23. Page, "Women and the Bicycle," 84.

24. Both quotes from R. L. Dickinson, "Bicycling for Women from the Standpoint of the Gynecologist," *American Journal of Obstetrics* 21 (1895): 25, cited in Haller and Haller, *Physician and Sexuality*, 185.

25. W. E. Fitch, "Bicycle-Riding: Its Moral Effect upon Young Girls and Its Relation to Diseases of Women," *Georgia Journal of Medicine and Surgery* 4 (1899): 155–156.

26. Arthur Conan Doyle, "The Adventure of the Solitary Cyclist," in *The Return of Sherlock Holmes* (New York: Berkley Books, 1963), 84; originally published in *Collier's Weekly*, December 26, 1903. The story is set in 1895.

27. George Napheys, A.M., M.D., *The Physical Life of Woman: Advice to the Maiden, Wife, and Mother* (Philadelphia: H. C. Watts, 1884), 40.

28. Mary Wood-Allen, *What a Young Girl Ought to Know* (Philadelphia: Vir, 1897), 106, cited in Haller and Haller, *Physician and Sexuality*.

29. See Napheys, *Physical Life of Woman*, 40.

30. Lina Beard and Adelia Beard, *The American Girl's Handy Book: How to Amuse Yourself and Others* (New York: Scribner's, 1898), 471.

31. Gormully & Jeffery Mfg. Co. Rambler Bicycles catalog, 1893, p. 23. Collection of the Winterthur Library.

32. Luther H. Porter, *Cycling for Health and Pleasure* (New York: Dodd, Mead, 1895), 52.

33. Automatic Cycle Seat, *Harper's Weekly*, April 11, 1896, p. 367.

34. The Sager Pneumatic Bicycle Seat, *Harper's Weekly*, April 11, 1896, p. 371.

35. *Dominion Monthly and Ontario Medical Journal* 7 (1896): 504, 11; (1898): 28, 30, cited in Patricia Vertinsky, *The Eternally Wounded Woman: Women, Doctors and Exercise in the Late Nineteenth Century* (Manchester: Manchester University Press, 1990), 75.

36. "Psyche" quoted in Porter, *Cycling for Health and Pleasure*, 18–20.

37. Rambler ad, *The Century*, June 1894, ad p. 38.

38. See, for example, Henry Garrigues, "Woman and the Bicycle," *The Forum*, January 1896, pp. 576–587. Garrigues, a doctor, proclaims that, by bicycling, a woman "far from diminishing her fitness for this supreme act in her life [childbirth], actually renders herself more capable of meeting the ordeal" (582).

39. Willard, *How I Learned*, 43.

40. Columbia Bicycles ad, *Harper's Weekly*, July 25, 1896, p. 38.

41. Merington, "Woman and the Bicycle," 703.

42. For more discussion on this, see Linda Gordon, *Women's Body, Women's Right: Birth Control in America* (New York: Penguin, 1990), 133–155.

43. Americanness was assumed to be white, as well as native-born. Although one bicycle historian notes that a homemade bicycle devised by a young black man who "according to *Cycle* magazine could not buy, rent, or borrow a bicycle" caused a stir in the mid-1890s (James Wagenvoord, *Bikes and Riders* [New York: Van Nostrand Reinhold, 1972], 89), the presence of such an anomalous piece of exotica only consolidated the notion that "real" bicycling was a white middle-class pursuit and that other people taking it up could do so only in grotesque imitation. African-Americans did ride bicycles, though they are absent from magazine imagery of riding. In April 1893 *Outing* magazine reported that the League of American Wheelmen convention had been embroiled in a fight ending only with the narrow defeat of an amendment barring black people from membership; the controversy was expected to continue: the league, begun by manufacturer Albert Pope, campaigned for better roads, organized races and clubs, arranged for discounts at inns, and did much to shape the practice and image of bicycling (ad p. 10).

In the only story seen for this study in which a black cyclist appears, he is a servant accompanying a girls' tricycling club on its tour. E. Vinton Blake, "The Girls' Tricycle Club and Its Run Down the Cape," *St. Nicholas*, May 1886, pp. 494–500.

44. See Christopher Wilson, *The Labor of Words: Literary Professionalism in the Progressive Era* (Athens: University of Georgia Press, 1985).

45. Presbrey, *The History and Development of Advertising*, 363.

46. Cleveland Moffett, "Great Business Enterprises: The Marvels of Bicycle Making," *McClure's*, February/May 1897, 50 pages paginated separately from both editorial and advertising sections. Others in this series of what would now be called advertorials—one for shoes and another for pianos—are more clearly flagged as ads. Pope's twelve-page condensation of the advertising articles in a booklet calls itself a reprint of a *McClure's* article. Cleveland Moffett, *How a Bicycle Is Made: One of America's Great Industries*, n.d., Bella Landauer Collection, New-York Historical Society, box: Bicycles 2.

47. Amy Kaplan, *The Social Construction of American Realism* (Chicago: University of Chicago Press, 1988), 10.

48. Joseph Bishop, "Social and Economic Influence of the Bicycle," *Forum*, August 1896, pp. 680–689.

49. See Maude Cooke, *Social Etiquette, or Manners and Customs of Polite Society* (n.p.: 1899), 343–344. It may be that the bicycle simply extended to the genteel middle class the upper-class acceptance of a young unmarried woman's horseback riding alone with a young man. This may have been acceptable, Louis Auchincloss suggests, because outdoor activities were seen as inherently wholesome. Auchincloss's commentary in Florence Adele Sloan, *Maverick in Mauve: The Diary of a Romantic Age* (Garden City, N.Y.: Doubleday, 1983), 126.

50. This may have been a specifically American notion, at least as it appears in fic-

tion. In H. G. Wells's *The Wheels of Chance* (London: J. M. Dent, 1896), bicycling allows a draper's assistant to be mistaken for a genteel educated man and to get entangled in a love plot. Stephen Kern reports in *The Culture of Time and Space, 1880–1918* (Cambridge: Harvard University Press, 1983) that in Maurice Leblanc's *Voici des ailes!* (Paris: P. Ollendorff, 1898) a bicycle trip sets off a rash of spouse swapping.

51. Flora Lincoln Comstock, "Rosalind Awheel," *Godey's Magazine*, April 1896, pp. 388–393.

52. *Godey's Magazine* in this period was no longer the genteel mainstay of women's magazines it had been as *Godey's Lady's Book*, but was attempting—ultimately unsuccessfully—to shape a more commercial identity, as it dropped its price to ten cents in 1894. See Frank Luther Mott, *A History of American Magazines, 1885–1905* (Cambridge: Belknap Press of Harvard University Press, 1957), 5, 87, and John Tebbel and Mary Ellen Zuckerman, *The Magazine in America 1741–1990* (New York: Oxford University Press, 1991), 36.

53. Adelaide Rouse, "The Story of a Story," *Munsey's*, October 1896, p. 47.

54. Bishop, "Social and Economic Influence of the Bicycle," 684.

55. Bernice Rogers, "The Reverend Abiel—Convert," *The Household*, November 1896, p. 6.

56. Harry St. Maur, "To Hymen on a Wheel," *The Home Magazine*, August 1897.

57. This and following quotes from John Seymour Wood, "A Dangerous Sidepath: A Story of the Wheel," *Outing*, June 1893, pp. 209–214.

58. Ibid.

59. For a discussion of a children's story in *St. Nicholas* magazine that uses some of the same elements (Gabrielle Jackson's "The Colburn Prize"), see Garvey, "Commercial Fiction," 339–342.

60. Charles H. Day, "Willing Wheeler's Wheeling: A Bicycle Story," *Home Magazine*, August 1897, p. 137.

61. Kate Chopin, "The Unexpected," in *A Vocation and a Voice: Stories* (New York: Penguin, 1991), 90–91. Originally published in *Vogue*, September 18, 1895.

62. Ibid., 93.

63. Emily Toth, *Kate Chopin* (New York: Morrow, 1990), 253, 279.

64. Mott, *History of American Magazines*, 759.

65. Willa Cather, "Tommy, the Unsentimental," in Joan Nestle and Naomi Holoch, eds., *Women on Women: An Anthology of American Lesbian Short Fiction* (New York: New American Library, 1990), 7–16. Originally in *Home Monthly* (Pittsburgh) 6 (August 1896).

The significance of the fact that I cite canonical writers who subvert the formula is *not* that greatness breaks the mold, but rather that although these stories appeared in relatively obscure magazines, the writers' later fame, and still later feminist scholarly interest in them, made the stories available in modern collections. Counterformulaic stories may have been published in less commercial magazines by writers who stayed obscure as well.

66. See Virginia Niles Leeds, "A Coast and a Capture: A Bicycling Story," *McClures*, July 1896, pp. 122–126.

67. All preceding quotes from Cather, "Tommy, the Unsentimental."

68. Ibid.

69. Setting the story in the somewhat exotic rural West, incidentally at a considerable distance from the main market for bicycles, may have made such a vision of bicycling less threatening. Poor roads in rural regions, as well as the depression of 1893 that cut into farmers' incomes, meant that few people living in rural areas owned bicycles.

70. Sharon O'Brien, *Willa Cather: The Emerging Voice* (New York: Oxford University Press, 1987), 227. O'Brien sees Cather's subversion here as the substitution of a male plot for a female one, a switch from romance to adventure story.

71. Annie Nathan Meyer, *It's Been Fun* (New York: Henry Schuman, 1951), 5.

72. See Virginia Scharff, *Taking the Wheel: Women and the Coming of the Motor Age* (New York: Free Press, 1991).

73. Among the historians who give this date are Robert A. Smith, *A Social History of the Bicycle: Its Early Life and Times in America* (New York: American Heritage, 1972).

Chapter 5. Rewriting Mrs. Consumer: Class, Gender, and Consumption

1. Amelia Barr, "Have Women Found New Weapons?" *Ladies' Home Journal*, June 1894, p. 4.

2. Preceding quotes from Edward Bok, *The Americanization of Edward Bok: The Autobiography of a Dutch Boy Fifty Years After* (New York: Scribner's, 1920), 11.

3. Ibid., 62–63.

4. See for example Gloria Steinem, "Sex, Lies and Advertising," *Ms.*, July/August 1990, pp. 18–28. This article, in the first issue of the advertising-free *Ms.*, discussed ways in which advertisers in women's magazines have directed editorial content to a focus on household and fashion concerns, and against, say, workplace or international issues.

5. Frank Luther Mott in *A History of American Magazines, 1885–1905* (Cambridge: Belknap Press of Harvard University Press, 1957) gives the circulation of *People's Literary Companion*, the first of these papers, as half a million as early as 1870, while *Comfort* was over one million in 1895.

6. Samuel Sawyer in *Secrets of the Mail Order Trade* (Waterville, Maine: Sawyer, 1900), points to one distinction between mail-order journals and other magazines: the mail-order journal's sales were entirely from subscriptions (typically a dollar or less a year), not individual copy sales, so that circulation would not respond to an individual issue's contents. He defines the class to which they were meant to appeal:

> They are cheaply printed on inexpensive paper and the literary matter in them is usually light fiction, interesting sketches, household column, 'little folks' corner,' 'chats with correspondents,' puzzles, agricultural hints and so forth. This kind of literature is attractive to people of the middle class [sic], and finds particular favor among those living in the small towns, farming and mining districts, as well as a portion of the city dwellers, such as mechanics and other workers' families. (113)

Although his use of the term *middle class* for farming and mining households is unusual, he is more specific about what he means the term to exclude. The contents "would hardly appeal to a banker, lawyer or railroad president" (115). They have readers, nonetheless, who are like the heroes and heroines of the stories in the mail order journals, "whose earnings are not sufficient to enable them to live like lords, but who nevertheless exist in their own homely way. These people eat and wear clothes. They have ailments requiring medicines, use soap, enjoy amusements, carry watches, also have numerous other habits and necessities. They are open to propositions to supply them with the commodities they require" (116).

7. *Ladies' World* not only sold commodities itself, but at least for a time offered its readers a "shopping agency." Presuming that its readers were rural, it asked in an ad announcing this service, "What do you need from the city?" (November 1889, p. 4). The

service relied on the readers' trust in the magazine and their sense of belonging to a community. The initial announcement traded on the reader's familiarity with the name of the publisher's wife, listed in 1904 as managing editor, and asserted her qualifications in the shopping sphere:

> The *Ladies' World* Agency will be conducted personally by Mrs. Myra D. Moore, who is well known to our readers, having been connected with the *Ladies' World* . . . for the past ten years. . . .
>
> She is a housekeeper, a mother and a lady of good taste, as shown by the things she gathers about her in her own home, and will make the wants of each customer her own for the time being, and endeavor to select exactly as if for her own use or wear.
>
> It is no new task she undertakes. For many years she has done shopping for out-of-town friends to the amount of hundreds of dollars, and has given in every instance the utmost satisfaction. (September 1889, p. 8)

Shopping is presented as a task which one might wish to delegate to someone else, rather than a pleasure. While it requires skill and taste, that taste is apparently universal rather than an expression of individuality: Myra Moore promises to select what *she* would prefer, not to express the customer's taste through purchases. This orientation toward shopping may not in fact have matched that of the *Ladies' World* readers: ads for the service don't appear past 1889.

8. In the April 1890 issue of *Ladies' World*, for example, twenty-eight ads offered either agencies for a product or a premium for selling it to neighbors. A number of these don't tell what the product is, but state, for example, "You can make a large sum of money at work for us in your own locality." Ad for True & Co., Augusta, Maine, p. 6.

9. Mott, *History of American Magazines* 360. I have been unable to locate any information about the authors whose works I have cited from *Ladies' World*. This may mean that *Ladies' World* attracted authors who were relative amateurs and who published a few items in periodicals only; it could also mean that the names were pseudonyms for an editor or author who wrote much of the magazine. (The exception is Mabael Gifford, who also wrote for *Woman's Home Companion*.) The magazine's editor, Frances Elizabeth Fryatt, had been a newspaper journalist and contributor to such magazines as *Harper's*, *The Independent*, *The Churchman*, *Art Age*, *Harper's Young People*, and *Wide Awake* before becoming editor-in-chief of *Ladies' World*, where she was acknowledged to have conducted eight of its departments, writing "all of the editorials and most of the technical articles." Whether this writing included fiction is not clear. "Frances Elizabeth Fryatt," in Frances Willard and Mary A. Livermore, eds., *American Women* (New York: Mast & Kirkpatrick, 1897). Myra Moore, too, was said to have written many unsigned pieces in *Ladies' World*. Marjorie Moore Butterworth, *Quawksnest: Its Memories and Those Who Made Them Possible, 1896–1979* (privately published, 1992), 18. I am indebted to Tracy Moore for showing me this family history. On the other hand, the publisher of *Ladies' World* asserted in a 1904 profile that its "high-grade fiction compares favorably with that published in any of the leading magazines" and indicated that it considered fiction an important feature of the magazine. "The Modern Magazines," *Judicious Advertising and Advertising Experience*, May 1904, p. 65.

10. Preceding quotes this paragraph from Velma Caldwell-Melville, "One Woman's Way," *Ladies' World*, August 1890, p. 15.

11. Adella F. Veazie, "The Rebellion of Reuel's Wife," *Ladies' World*, serialized March–May 1904, March, p. 4.

12. The author also establishes that Mazie's parceling out her responsibilities,

including hiring help for some, is in line with the same division of labor already accepted in men's work, including farm work. As Mazie explains that she plans to keep a hired girl and pay her out of her earnings, she and Reuel have this exchange:

> "I have learned that I am utterly unqualified and unfitted to do housework, and the more I try, the worse I hate it. I have tried faithfully, but I really think I'd rather die than look forward to seeing my whole life spent in the kitchen."
>
> Here Reuel cast on her a look of astonishment.
>
> "But the housework is *yours*," he said in a tone of displeasure. "You don't mean that you intend to shirk it altogether?" interrogatively.
>
> "So is your horse and your wagon and your barn yours," she retorted, "but you didn't shoe your horse yesterday; you took it to the blacksmith. You didn't make a new tongue to your wagon when the old one gave out; the wheelwright got that job, and instead of building your barn yourself, you hired carpenters to do the work." (Veazie, "The Rebellion," April 1902, p. 15)

13. In *The Secret of a Happy Home*, Marion Harland (the pseudonym of Virginia Terhune) uses "John" as her generic name for a husband and proclaims "John is not John until he is married. He assumes the sobriquet at the altar as truly as his bride takes the title of 'Mistress' or 'Madame'" (New York: Christian Herald, 1896), 19.

14. M. Vaughn, "John's Wife," *Ladies' World*, October 1891, 3.

15. Emily Hayes, *Cousin John's Extravagant Wife* (Diamond Dyes, 1898), unpaginated. In the Thelma Mendsen Collection, pamphlet file, Winterthur Library. "Emily Hayes" may not have been a pseudonym. A writer of that name published at least one long letter in 1883 in the monthly periodical *The Household*. The letter advocated that rural young people form reading clubs to fill up the long winter evenings with study and recreation—a rather different use of time than dyeing clothes, though both projects recognized that rural people might have free time to put to use. Emily Hayes, letter from *The Household*, February 1883; reprinted in Norton Juster, *So Sweet to Labor: Rural Women in America 1865–1895* (New York: Viking, 1979), 175.

16. Ibid.

17. Ibid.

18. Women found nonlegitimate means to acquire goods as well, especially in the face of what they experienced as the overwhelming temptations of the department store. If the magazine as a two-dimensional version of the department store presented some of the same temptation, if not the same opportunity, the stories suggested a way to master them by first acquiring the money and then the goods. For more on department store theft in this period, see Elaine Abelson, *When Ladies Go A'Thievin': Middle-Class Shoplifters and the Victorian Department Store* (New York: Oxford University Press, 1989).

19. All preceding quotes from Bertha Ashton, "Aunt Crawford's Wise Will; or Perseverance Conquers All Things," *Ladies' World*, March 1892, p. 15. The story ran on the page labeled "Mother's Department," which included advice to mothers and stories for children.

20. Barr, "Have Women Found New Weapons?" p. 4.

21. Isabel A. Mallon, "Dressing Neatly at Breakfast," *Ladies' Home Journal*, March 1893, p. 19.

22. George Hodges, D.D., "The Business of Being a Wife," *Ladies' Home Journal*, January 1906, p. 19.

23. For more on the changes in expectations of women as mothers, see Maxine Margolis, *Mothers and Such: Views of American Women and Why They Changed* (Berkeley:

University of California Press, 1984), especially her discussion in chapter 2 of rearing middle-class "quality children" and pages 110–112, which touch on the middle-class woman's work of creating a haven for her husband at home. See also Ruth Schwartz Cowan, *More Work for Mother: The Ironies of Household Technology from the Open Hearth to the Microwave* (New York: Basic, 1983); Christina Hardyment, *Dream Babies: Three Centuries of Good Advice on Child Care* (New York: Harper & Row, 1983); Glenna Matthews, *"Just a Housewife": The Rise and Fall of Domesticity in America* (New York: Oxford University Press, 1987); and Susan Strasser, *Never Done: A History of American Housework* (New York: Pantheon, 1982).

24. In a 1902 novel, Rouse took a very different approach to female authorship. The heroine of her *Under My Own Roof* (New York: Funk and Wagnalls, 1902) is an unmarried successful writer of popular novels and stories who has not only a room of her own but an entire house built on the proceeds of her writing. When she marries a (more serious) fellow writer, they carefully adapt her house to provide writing rooms for both of them. One reviewer complained that the heroine earned money from her writing *too* easily and would thereby mislead readers: "All the adventures which befall Honoria are possible and within the defined realms of romance. We only wish that the ready-money business obtainable in the way Honoria shoveled it in were credible. It may be that Adelaide L. Rouse swims in a Pactolus-like flood, but we are afraid that scores of young women, and belike old ones, will be trying to go to sea on floats of paper, using pens for oars and suffer shipwreck." The reviewer's metaphor of a mythological destructive flood perhaps hints at the fear that such inspired writers might flood the market. (Quotes are from unsigned review, *The New York Times*, September 6, 1902, p. 603.) As the author of at least ten books and as editor of Frances Trollope's *Domestic Manners of the Americans*, Rouse, who was "editorially connected with publishing houses" was very much a professional writer. *Who Was Who in America*, vol. 1, *1897–1942* (Chicago: Marquis, 1966), 1061.

25. Preceding quotes from Marguerite Tracy, "The Unhonored Profession," *Munsey's*, March 1901, pp. 941–942.

26. Preceding quotes from Bessie Chandler, "The Woman's Edition," *Ladies' Home Journal*, May 1896, pp. 5–6.

27. The appearance of this detail in a story in the *Ladies' Home Journal*, incidentally, set the women's paper on a lower moral plane than the *Journal*, which publicized its refusal of advertising for patent medicines; *Journal* editor Edward Bok noted in his autobiography that he "got the Women's Christian Temperance Union into action against the [other] periodicals for publishing advertisements of medicines containing as high as forty percent alcohol." Bok, *Americanization*, 340–341.

28. Bessie Chandler, "A Woman Who Failed," *A Woman Who Failed and Other Stories* (Boston: Roberts Brothers, 1893), 20.

29. E. M. Halliday, "Mrs. Medlicott," *Munsey's*, December 1893, pp. 261–263.

30. Mabel Gifford, "Maria's So Thoughtless: A New-Year Story," *Ladies' World*, January 1892, pp. 2–3.

31. Matthew White Jr., "In the Shadow of Success," *The Puritan*, January 1897, p. 28. White was well ensconced in the literary profession as the drama editor of *Munsey's* for twenty-eight years, the editor of Munsey's children's magazine *Argosy* for forty years, and the author of at least eight children's and adult books.

32. Ibid., 30.

33. Henry Ferris, [Note from the editor] *Housekeeper's Weekly*, May 17, 1890, p. 7. The masthead lists Ferris as the editor and manager, Bertha A. Winkler as assistant manager, and Louise B. Edwards as assistant editor.

34. Janice Radway, *Reading the Romance: Women, Patriarchy, and Popular Literature* (Chapel Hill: University of North Carolina Press, 1984).

35. See Mary Ellen Waller, "Popular Women's Magazines, 1890–1917" (Ph.D. diss., Columbia University, 1987), 150, 154.

36. See for example Gabriel Kolko, *The Triumph of Conservatism: A Reinterpretation of American History, 1900–1916* (New York: Free Press, 1963).

37. Unsigned, "Deceptive Ads," *Profitable Advertising and Art in Advertising*, October 1899, p. 336.

38. Steinem, "Sex, Lies and Advertising," 26, 27. For more on "advertising reciprocity" in present-day magazines, see Ellen McCracken, *Decoding Women's Magazines: From Mademoiselle to Ms.* (New York: St. Martin's, 1993), particularly the chapters "The Cover" and "Covert Advertisements."

39. Steinem, "Sex, Lies and Advertising," 26.

40. Waller, "Popular Women's Magazines," 151.

41. The focus on obtaining the work of prominent authors also meant that the contents of an individual issue became important, differentiating the *Journal* from the subscription-driven sales of *Ladies' World* and other mail-order magazines.

42. Edward Bok, "At Home with the Editor," *Ladies Home Journal*, October 1893, p. 16.

43. Ibid.

44. Ros Ballaster, Margaret Beetham, Elizabeth Frazer, and Sandra Hebron, *Women's Worlds: Ideology, Femininity and the Woman's Magazine* (Houndmills and London: Macmillan, 1991), 98–99, points to British nineteenth century domestic women's magazines as similarly defining women as outside the public realm.

45. Edward Bok, "At Home with the Editor" *Ladies' Home Journal*, January 1894, p. 12.

46. Charles LeGuin, ed., *A Home-Concealed Woman: The Diaries of Magnolia Wynn LeGuin* (Athens: University of Georgia Press, 1990), 15, 325, 33–314, 92–93, 192.

47. Bok envisioned his reader as a small-town woman, specifically not a farm woman, and enforced that vision through his editorial choices. Editorial correspondence with the author of an article on "An Ideal Farm Garden" finds Bok's assistant requesting that the author scale down the size of the proposed garden: "Mr. Bok . . . maintains that he probably has a clearer idea than you can possibly have, as to what would probably please the majority of readers of the Journal. He . . . thinks that it would be much better for our purposes if we were simply to describe a garden such as would be suitable for those who live in the suburbs of large cities, or in the small country towns, where people have even less than an acre of land for their houses, gardens, and all that." Letter to Miss Lucy Cleaver McElroy, January 30, 1900, Curtis Publishing Co. records, Special Collections, Van Pelt Library, University of Pennsylvania.

Helen Damon-Moore, in her study of the *Ladies' Home Journal* and the *Saturday Evening Post*, also noted that the *Journal,* by the early twentieth century, was mainly directed at families with "moderate middle-class incomes from $1200 to $2500," with some attention to those with incomes from $2500 to $5000. Helen Damon-Moore, *Magazines for the Millions: Gender and Commerce in the* Ladies' Home Journal *and the* Saturday Evening Post *1880–1910* (Albany: State University of New York Press, 1994), 73.

48. Stories of the rich woman with a poor but hardworking suitor were common in other magazines as well. See for example, Thomas Baily Aldrich, "His Dying Words," *Scribner's*, August 1893, pp. 204–211, and Harrison Robertson, "The Rich Miss Girard," *Scribner's*, September 1893, pp. 390–393.

49. Preceding quotes from Juliet Wilbor Tompkins, "His Dutch-Treat Wife," *Ladies' Home Journal*, June 1905, p. 12.

50. Helen Damon-Moore's study of the *Journal* takes note of other ways in which the ads seem to conflict with the editorial stance. She locates the point of agreement in "the themes that buying and consuming represented a natural way of life, that efficiency was an ideal to be sought above all others, and that intelligence equalled buying and using the right products" (*Magazines for the Millions*, 106).

51. *Ladies' Home Journal*, January 1896, p. 41.

52. *Ladies' Home Journal*, December 1895, p. 31.

53. John Kendricks Bangs, "The Paradise Club," *Ladies' Home Journal*, December 1894, p. 9.

54. Edward Bok, "At Home," October 1893.

55. Mallon, "Dressing Neatly," 19.

56. Candace Wheeler, quoted in Eileen Boris, *Art and Labor: Ruskin, Morris, and the Craftsman Ideal in America* (Philadelphia: Temple University Press, 1986), 53–54.

57. For a fuller discussion of this theme in the writing of the period, see Garvey, "Commercial Fiction: Advertising and Magazines 1880s to 1910s" (Ph.D. diss., University of Pennsylvania, 1992).

58. Annie L. Hannah, "Miss Deborah's Investment: A Thanksgiving Story," date and publication unknown; in scrapbook of stories evidently clipped from news or story papers directed at a rural audience beginning in the late 1880s. From the collection of the author.

59. In a rare story in which shopping is the featured action, the match between goods and buyer is monogamous (or perhaps serially monogamous): Katherine hesitates too long and loses the chance to buy a collar that exactly matches her shirtwaist. Her friends and relatives send her a total of eighty-three collars, but none matches. Months of effort and correspondence later, she finds the collar's buyer, who sends her the collar, saying it didn't match her own shirtwaist. A. Carroll Watson Rankin, "The Quest of a Nile-Green Collar," *St. Nicholas*, August 1905, pp. 889–891.

60. As Susan Strasser explains, advertisers encouraged consumers to buy the advertised product instead of the generic item the grocer might promote. *Satisfaction Guaranteed: The Making of the American Mass Market* (New York: Pantheon, 1989) 261. To do this, ads valorized choice making with scenarios that showed women refusing the "wrong" product or suffering dire consequences from buying it.

61. Judith Williamson, *Consuming Passions: The Dynamics of Popular Culture* (London: Marion Boyars, 1986), 6.

62. "A Substitutor's Obituary," *Judicious Advertising and Advertising Experience*, July 1903, p. 25.

63. Rachel Bowlby, *Just Looking: Consumer Culture in Dreiser, Gissing, and Zola* (New York: Methuen, 1985), 27.

64. Thomas Winthrop Hill, "Her Triumph," *Munsey's*, October 1893, p. 53.

65. It is possible to read such stories as advice to women on the dangers of marrying for money, an approach taken by Frances Cogan reading mid-nineteenth-century novels. Similar scenarios appear in the works Cogan cites: a woman must choose between a rich man she hates and a poor but rising young professional. Her sample finds the women marrying the rich men; however, in the later magazine stories, the woman often wisely chooses the rising young man. Cogan, *All-American Girl: The Ideal of Real Womanhood in Mid-Nineteenth Century America* (Athens: University of Georgia Press, 1989), 110.

66. Harriet Caryl Cox, "His Mother," *Munsey's*, April 1898, p. 134.

67. Ibid.

68. Preceding quotes from Anne O'Hagan, "Rhodora, Advance Agent of the Better Day," *Good Housekeeping*, May 1909, pp. 557–567.

69. Florence Kelley, "Aims and Principles of the Consumers' League," *American Journal of Sociology* 5, no. 3 (November 1899): 290; quoted in Joe Broderick, "The Discovery of the Consumer as a Social Force: The Consumers' Leagues and Their Reform Strategies, 1890–1900," unpublished paper presented at the Rutgers Center for Historical Analysis, May 5, 1992. Information on the Consumers' League is from Broderick.

70. O'Hagan, "Rhodora," 567.

71. Preceding quotes from Lulu Judson, "A Girl's Way," in "Storiettes" section, *Munsey's*, November 1896, pp. 205–207.

72. Unsigned, "Deceptive Ads," p. 336.

73. F. W. Leavitt, "Who Pays the Publisher: Subscriber or Advertiser?" *Mahin's Magazine*, August 1902, p. 23.

Chapter 6. "Men Who Advertise": Ad Readers and Ad Writers

The title of this chapter is from a promotional book by one of the first advertising agency heads: George Rowell, *The Men Who Advertise: An Account of Successful Advertisers Together with Hints on the Method of Advertising* (New York: Nelson Chesman, 1870).

1. Unsigned editorial, "Ethics and Esthetics of Advertising," *The Independent* 55, no. 2770 (January 2, 1902): 52–53.

2. Noting the dubiousness of much research on how ads influence consumer sales, Michael Schudson suggests that all ads are ultimately directed at stockholders rather than consumers, reassuring them of the company's importance. Similarly, while these articles seem to be aimed at readers, they may have more effectively reinforced for advertisers the notion that ads were valuable. Michael Schudson, *Advertising: The Uneasy Persuasion: Its Dubious Impact on American Society* (New York: Basic Books, 1984), xiv.

3. Unsigned, "Magazine Advertising" in "Editorial Comment," *Current Literature*, January 1903, p. 9.

4. Ibid.

5. Ibid.

6. E. C. Patterson, "Advertising Bulletin Number 4: The Cost of Advertising," *Collier's*, May 22, 1909, p. 5.

7. Robert F. Bishop, "Literature and Business," *Art in Advertising*, October 1890, pp. 146–147.

8. Margaret Holmes Bates, "Is Advertising a Science?" *Judicious Advertising and Advertising Experience*, May 1906, p. 67. Advertising might literally be more timely than other contents of a magazine. *St. Nicholas's* printing schedule, for example, meant that material could appear in the advertising section three to four weeks more rapidly than in the body of the magazine. Correspondence: William Fayal Clarke, editor, to Don M. Parker, February 13, 1914, The Century Collection, Box 74, New York Public Library.

9. "The New Shopping: Enthusiastic Letters from Readers Who Have Made Excursions into 'Advertising Land,'" *Good Housekeeping*, January 1913, pp. 130–131.

10. Frank Munsey Jr., "The Publisher's Desk," *Munsey's*, July 1897, p. 638. Munsey notes that his reply repeats something from a column over a year before. Nathaniel Fowler uses virtually the same terms in his advertising advice book: "The advertising pages of the great magazines reflect public opinion; they are announcements of necessity and convenience. They mirror the business world, and are really as philanthropic as

they are businesslike." Fowler, *Fowler's Publicity: An Encyclopedia of Advertising and Print-ing, and All That Pertains to the Public-Seeking Side of Business* (New York: Publicity Pub-lishing, 1897), 375.

11. Munsey, "The Publisher's Desk," 638.

12. Samuel Hopkins Adams, "The New World of Trade," *Collier's,* May 22, 1909, p. 14.

13. Ibid., 15.

14. Annie Nathan Meyer, *Bookman* March 1903, cited in "Among the February Magazines," *Judicious Advertising and Advertising Experience* 1, no. 5 (March 1903): 23.

15. "Literary Notes," *Art in Advertising,* November 1890, p. 163.

16. See, for example, "Our New Department," *Art in Advertising,* November 1890, p. 170, complaining of the reciprocal promotion engaged in by *Harper's* and *The Century.*

17. Finley Peter Dunne, "Mr. Dooley on the Magazines," *American Magazine,* Sep-tember 1909, pp. 539–542.

18. Robert F. Bishop, "Literature and Business," *Art in Advertising,* October 1890, pp. 146–147.

19. Bates, "Is Advertising a Science?" 67.

20. *Mrs. John Doe: A book wherein for the first time an attempt is made to determine woman's share in the purchasing power of the nation* (New York: Butterick Publishing, 1918), 27–28. Butterick published *The Delineator,* a major women's magazine, which claimed a circulation of half a million in 1895, four years after the *Ladies' Home Journal* did, and one and a half million in 1906, according to Frank Luther Mott, *A History of American Magazines, 1885–1905* (Cambridge: Belknap Press of the Harvard University Press, 1957), 16.

21. Frank M. Low, "How a Men's Store Was Advertised," in *Advertising: Selling Points and Copy Writing* (reprint of *The Library of Business Practice,* vol. 6, 1914; Chicago: A. W. Shaw, 1917), 158.

22. Nancy Finlay, "The Girl in the Poster: Images of Women Reading in Turn-of-the-Century America," unpublished paper, 1991.

23. Ad for "'Sans Souci' Hammock, No. 2," *Ladies' Home Journal,* July 1892, p. 35.

24. In asserting that the women's magazine is a trade journal, the *Journal's* pam-phlet addressed to advertisers suggests that the home is still a realm of production, and that it works along a social Darwinist model: "The home is her factory. There raw mate-rials are being converted into finished products, flour into pastry, cloth into clothes. There she competes with other men's wives in the dressing of her children, in the dain-ties on her table, in the tasty arrangement of her living-room." *Selling Forces* (Philadel-phia: Curtis, 1913), 230.

A 1900 commentator distinguished between magazines and "household publica-tions," a category that would probably include the *Journal* as well as *Ladies' World,* and perhaps *Comfort* and the *People's Literary Companion.* She called these household publi-cations a woman's "trade paper. . . . She may read the magazine, but she studies her household paper, just as the merchant studies his trade journal." Marion J. McKenzie, "Household Publications, from a Woman's Standpoint," *Profitable Advertising and Art in Advertising,* April 1900, p. 791.

25. Fowler, *Fowler's Publicity,* 374. Because advertisements appeared in a separately paginated section in the back and front of the magazine, they could be printed separately and only the ad pages trimmed before binding, rendering them more accessible than the editorial matter. The pages in this central section, like those of expensive editions of books, were left uncut for the reader to slice open. (This distinction has often been obliterated in magazines bound together for library use, where pages have been uniformly trimmed.)

26. Samuel Hopkins Adams, "The New World of Trade," *Collier's*, May 22, 1909, p. 13.

27. Hrolf Wisby, "Modern Advertising Methods," *The Independent*, February 1904, p. 261.

28. "Fulkerson" (pseud.), "Causerie," column, *Art in Advertising*, March 1890, p. 5. The character of Fulkerson from *A Hazard of New Fortunes* is given a somewhat altered career here, not simply advertising his own magazine as he does in Howells's novel, but taking an interest in advertising within magazines. (That the use of the character's name is deliberate is established by a page of extracts from Fulkerson's conversations in *Hazard*, "How to Run a Magazine: or, how Mr. Howells thinks it should be done." Howells's publisher soon sensibly followed up this plug with a full page ad for *Hazard* in a subsequent issue of *Art in Advertising*.)

29. Beginning in the 1900s, ad trade journals spoke of women being responsible for at least 80 percent of purchases; some challenged the dogma and raised the figure to 90 or 95 percent. None cites evidence for any of the figures.

30. Wisby, "Modern Advertising Methods," 261.

31. Jackson Lears points to an 1840s version of "the triangular imagery of consumption" in which the peddler is the seducer and woman the prey; the advertiser later replaces the peddler in the seducer role. The cuckold in this scenario can be either the husband, who might be impoverished by his wife's extravagance, or the staid merchant, who cannot keep up with the attractions of the peddler or distant advertiser. Jackson Lears, *Fables of Abundance: A Cultural History of Advertising in America* (New York: Basic, 1994), 72.

32. *Selling Forces*, 54.

33. *Profitable Advertising*, February 1905, p. 975.

34. See Thomas Richards's provocative study, *The Commodity Culture of England: Advertising and Spectacle, 1851–1914* (Stanford: Stanford University Press, 1990), 246.

35. Fowler, *Fowler's Publicity*, 725.

36. *Selling Forces*, 229.

37. Rachel Bowlby, *Just Looking: Consumer Culture in Dreiser, Gissing, and Zola* (New York: Methuen, 1985), 19–20.

38. James Barrett Kirk, "A Midsummer Madness," *Profitable Advertising*, August 1899, pp. 169–170.

39. Ibid., 170.

40. Frank M. Low, "How a Men's Store Was Advertised," no editor, *Advertising: Selling Points and Copy Writing* (Chicago: A. W. Shaw, 1917), 158.

41. J. Angus MacDonald, *Successful Advertising: How to Accomplish It* (1902; reprint, Philadelphia: Lincoln Publishing, 1906), 243, under the heading "Advertising to Men."

42. Ibid., 241, under the heading "Advertising to Women."

43. Ibid.

44. Sherwin Cody, "Good English for Advertisers: IV. The Art of Writing Business-Winning Letters," *Profitable Advertising*, May 1904, p. 1287.

45. Working women's time is not discussed in these articles, suggesting either that their reading habits were seen as so gender-linked as to be undifferentiated from housewives' or that they were not considered a worthwhile market.

46. Coming out of the depression of the 1890s, farmers were seen as potential customers, and also as eager for reading matter. "The farmer does not subscribe for many papers, but he reads thoroughly those he does subscribe for," asserted one advertising trade journal. "Is the Trade of the Farmer Worth Cultivating?" *Mahin's*, April 1903, p. 49.

47. Preceding quotes from Oscar Herzberg, "Why Women Read Advertisements," *Mahin's*, April 1903, pp. 16–17.

48. Witt K. Cochrane, "Can You Write Good English?" *Mahin's*, April 1903, p. 47.

49. R. B., "Old-Fashioned Advertising," *Living Age*, July 16, 1898, p. 207. Reprinted from *The Gentleman's Magazine*, a British publication.

50. Frank Munsey Jr., "A Generation of Writers," in "The Publisher's Desk" column, *Munsey's*, July 1895, p. 438.

51. Frank Munsey Jr., "There Is Room for Men," in "Impressions by the Way" column, *Munsey's*, March 1896, p. 762.

52. Munsey calls for "us" to "begin to do men's work in fiction" in both "A Generation of Writers" in his "Publisher's Desk" column, *Munsey's*, July 1895, p. 438, and in "There Is Room for Men" in his "Impressions by the Way" column, March 1896, p. 762.

54. Frank Munsey Jr., "The Modern Advertisement," in "The Publisher's Desk" column, *Munsey's*, July 1895, p. 438.

55. Frank Munsey Jr., "Some Good Stories This Month," in "The Publisher's Desk" column, *Munsey's*, June 1895, p. 319.

56. Helen Woodward, *Through Many Windows* (New York: Harper, 1926), 158, 160–161. As a woman in an almost entirely male profession, Woodward saw herself as very different from the women who read the women's magazines in which her ads directed to women appeared.

57. Ellis Parker Butler, "The Adventure of the Poet," in *Perkins of Portland: Perkins the Great* (Boston: Herbert B. Turner, 1906), 77–94.

58. Ibid.

59. *Selling Forces*, 86.

Conclusion

1. Unsigned editorial, "Ethics and Esthetics of Advertising," *The Independent* 55, no. 2770 (January 2, 1902), 53.

2. Susannah James, *Love Over Gold* (London: Corgi, 1993), jacket front, back, and inside covers.

3. Leslie Savan's *The Sponsored Life: Ads, TV, and American Culture* (Philadelphia: Temple University Press, 1994) provides lively and insightful analysis of ads that use this technique.

Index